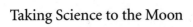

Taking Science to the Moon

New Series in NASA History

Roger D. Launius, Series Editor

Taking

Science

to the

Moon

Lunar Experiments and the Apollo Program

Donald A. Beattie

The Johns Hopkins University Press

Baltimore and London

© 2001 The Johns Hopkins University Press
All rights reserved. Published 2001
Printed in the United States of America on acid-free paper

9   8   7   6   5   4   3   2   1

The Johns Hopkins University Press
2715 North Charles Street
Baltimore, Maryland 21218-4363
www.press.jhu.edu

Library of Congress Cataloging-in-Publication Data
Beattie, Donald A.
Taking science to the moon: lunar experiments and the Apollo Program
/ Donald A. Beattie.
p. cm. — (New series in NASA history)
Includes bibliographical references and index.
ISBN 0-8018-6599-9
1. Project Apollo (U.S.) 2. Apollo Lunar Surface Experiments Package.
3. Moon—Exploration—Instruments. I. Title. II. Series.
TL789.8.U6 A5186 2001
629.45'4'0973—dc21   00-010272

A catalog record for this book is available from the British Library.

Frontispiece: Night launch of the final Apollo mission (*Apollo 17*)
as viewed from Kennedy Space Center, December 7, 1972. (NASA S72-55070)

Contents

Contents

Preface

The technical achievements that permitted the National Aeronautics and Space Administration (NASA), other government agencies, and their contractors to fulfill President John F. Kennedy's promise of "landing a man on the moon and returning him safely" have often been described. Most previous authors have included anecdotes that enhance our appreciation of how Project Apollo was successfully accomplished, although many are retold at second or third hand. Several movies such as *The Right Stuff* and *Apollo 13* showed both true and fictional accounts of the spirit and engineering skills that characterized the entire project, focusing primarily on the major or well-known participants.

A story that has not been completely told, however, is how a small band of somewhat anonymous NASA staffers, allied with scientists inside and outside government, struggled to persuade the management of NASA to look beyond the initial *Apollo* landing and reap a scientific harvest from this historic undertaking. Here is that story as seen through the eyes of a participant based at NASA headquarters—a pack rat who kept many of the internal memos, reports, photos, and notes that document that ten-year struggle. It highlights the contributions of many of those who worked with me during the Apollo program. Some of them have received little public recognition for their efforts. I hope that this insider background will give readers a better understanding of the behind-the-scenes maneuvering that led to many of Project Apollo's scientific achievements, which have enriched our understanding not only of the Moon but, more important, of the small planet we call Earth.

Acknowledgments

Many people and organizations helped and encouraged me while I was writing this book, and they deserve credit. Although I had saved many boxes of material I collected during my Apollo days at NASA in anticipation of one day writing this story, I soon found this source material was insufficient. Calling old colleagues to ask if they had kept records was not very fruitful at first, but eventually I was successful.

The first person who agreed to share his records covering part of this period was Robert Fudali, who was on the Bellcomm staff during Apollo's early days. His material not only contributed to the accuracy of this story but served as a valuable reminder of some of the events that occurred during the formative years of Apollo science. I have quoted liberally from a few of Bob's colorful internal memos.

Gordon Swann, a friend, former colleague, and principal investigator who took part in the struggle to develop science payloads for Apollo, especially those aspects related to the astronauts' geological investigations, reviewed early drafts and provided many important comments and suggestions as well as a few of his famous anecdotes—some printable, some not. Gordon should be encouraged to one day write his account of Apollo.

Paul Lowman, who figures prominently in this story, was an invaluable source of material and a resource for clarifying many events. Paul is renowned among his NASA colleagues as a pack rat of the first degree: his office is so filled with reports and trivia that when you first enter it is hard to find his desk. However, his propensity for maintaining his archives has benefited many who have written about NASA's early days. He also reviewed the manuscript and offered many useful comments.

Acknowledgments

James Downey, Herman Gierow, Farouk El Baz, and Charles Weatherred reviewed drafts at various stages, and Jim spent many hours going through the files at the Marshall Space Flight Center library to select material relating to the early years of our post-Apollo work. Chuck Weatherred and Eugene Zaitzeff (both Bendix employees during Apollo) and Charles Spoelhop at Eastman Kodak also provided important background material from their files. My former colleagues Philip Culbertson, Richard Allenby, Edward Davin, Richard Green, George Esenwein, Alex Schwarzkopf, Saverio "Sonny" Morea, George Ulrich, Raymond Batson, William Muehlberger, Floyd Roberson, and John Bensko took the time to provide information and pictures and to confirm recollections now more than thirty years old. Hugh Neeson, a former Textron-Bell engineer, searched the archives of the Niagara Aerospace Museum to find rare artists' drawings of the lunar flying vehicle. Bruce Beattie, my son, became a fact finder after I moved from Maryland, following up on questions that could be answered by Washington sources.

The NASA headquarters history office, in particular Lee Saegesser (before he retired) and Roger Launius and his staff, helped me access the records still maintained in Washington. Glen Swanson, NASA Johnson Space Center (JSC) historian, provided key contacts at JSC, including Joseph Kosmo at the Flight Crew Support Division and Judith Allton in the lunar sample curator's office that allowed me to fill in a few blanks in my story. And most important Michael Gentry and David Sharron at the JSC Media Resource Center, who spent considerable time helping me select and acquire the photos and drawings in the book.

Roger Van Ghent, a colleague and fellow Floridian, advised me on the intricacies of using my computer to ease my writing load and also helped compile the index.

To all these people and the many colleagues and friends whose names do not appear, my sincere thanks for your help and encouragement from my first days at NASA until the present.

Finally, I thank Alice Bennett at the University of Chicago for editing and improving the manuscript and Bob Brugger, my editor at the Johns Hopkins University Press, for running interference and patiently guiding me through the publishing process. There is no substitute for an unflappable editor.

Abbreviations and Acronyms

AAP	Apollo Applications Program
AEC	Atomic Energy Commission
AES	Apollo Extension System
AFO	announcement of flight opportunities
ALSEP	Apollo Lunar Surface Experiments Package
ALSS	Apollo Logistics Support System
ASSB	Apollo Site Selection Board
CapCom	capsule communicator
CCG	Cold Cathode Gauge (experiment)
CM	command module
CSM	command and service module
EASEP	Early Apollo Scientific Experiments Package
EMU	extravehicular mobility unit
ESS	Emplaced Scientific Station
EVA	extravehicular activity
FOD	Flight Operations Directorate
FRR	flight readiness review
GLEP	Group for Lunar Exploration Planning
GSFC	Goddard Space Flight Center
JPL	Jet Propulsion Laboratory
JSC	Johnson Space Center
KSC	Kennedy Space Center
LEM	lunar excursion module
LESA	Lunar Exploration Systems for Apollo
LFV	lunar flying vehicle

LM	lunar module
LOR	Lunar orbit rendezvous
LRL	Lunar Receiving Laboratory
LRRR	Laser Ranging Retro-Reflector (experiment)
LRV	lunar roving vehicle
LSAPT	Lunar Sample Analysis Planning Team
LSPET	Lunar Sample Preliminary Examination Team
LSSM	local scientific survey module
MET	modularized equipment transporter
MOCR	Mission Operations Control Room
MOLAB	mobile laboratory
MSC	Manned Spacecraft Center
MSFC	Marshall Space Flight Center
MSFEB	Manned Space Flight Experiments Board
NASA	National Aeronautics and Space Administration
OMSF	Office of Manned Space Flight
OSS	Office of Space Science
OSSA	Office of Space Science and Applications
PI	principal investigator
PLSS	portable life-support system
RFP	request for proposal
RTG	radioisotope thermoelectric generator
SEP	Surface Electrical Properties (experiment)
SIDE	Suprathermal Ion Detector Experiment
SIM	scientific instrument module
SIVB	Third stage of *Saturn V*
SM	Service Module
SSB	Space Science Board (also Source Selection Board)
SSR	Science Support Room (also Staff Support Room)
SSSC	Space Science Steering Committee
SWS	Solar Wind Spectrometer (experiment)
USGS	United States Geological Survey
VOA	Voice of America

Introduction

Anchored to its launch pad on the morning of July 16, 1969, and scheduled to launch *Apollo 11* on our first attempt to land men on the Moon, the fully fueled *Saturn V* launch vehicle weighed over six million pounds. From the nozzles at the base of the giant S-1C first stage to the top of the solid rocket–propelled escape tower, it measured 363 feet. In 1962, one year after President Kennedy had given the go-ahead for Project Apollo, the critical decisions had been made on how to execute his difficult challenge. *Saturn V,* with its multiple stages, was the key to reaching the goal, the product of seven years of effort by hundreds of thousands of government and contract workers.

The original planning in 1960 and 1961 centered on building a huge rocket to launch a spacecraft directly from Earth to the lunar surface, followed by a direct return home. The mission design finally selected was very different. It required a smaller, but still very large, multistage rocket to launch three astronauts into a low Earth orbit and then send them on to the Moon in a spacecraft that combined command and logistics modules with a lunar lander. On arriving at the Moon, these combined spacecraft would be parked in a low lunar orbit. The lunar lander, a two-stage (descent and ascent stages) two-man spacecraft, would then separate and go to the lunar surface. The command and service module, with the third astronaut on board, would remain in lunar orbit to rendezvous and link up with the astronauts when they returned from the Moon's surface. After the astronauts who had landed on the Moon transferred back to the command module, they would jettison the lunar lander ascent stage, and all three would leave lunar orbit and return to Earth in the command module for an ocean recovery.

Lunar orbit rendezvous (LOR) was the unique feature of the mission design that allowed NASA to reduce the size of the initial launch vehicle. An LOR flight profile required the development of a new, powerful rocket (*Saturn V*) and the design and fabrication of two complex spacecraft that would perform a series of difficult and potentially dangerous space maneuvers never before attempted. But a manned lunar landing designed around LOR was sold to NASA management as the quickest, least risky, and lowest-cost way to carry out the president's mandate. The LOR decision fixed the broad architecture of the mission and defined the parameters within which the scientific community would have to work when NASA finally determined what scientific activities were appropriate for future Apollo astronauts to carry out. (How NASA decided to adopt LOR, in a behind-the-scenes debate, has been covered in some detail in several of the references cited.)

Because the president's mandate did not require that any specific tasks be accomplished once the astronauts arrived on the Moon, the initial spacecraft design did not include weight or storage allowances for scientific payloads. Somewhere, somehow, amid the six million pounds and 363 feet, we would have to squeeze in a science payload. The earliest thinking was, "We'll land, take a few photographs, pick up a few rocks, and take off as soon as possible." The need to do much more was not considered in the planning. For many NASA engineers and managers the lunar landing was a one-shot affair. After the first successful landing, NASA would pack up its rockets and do something else. Why take any more chances with the astronauts' lives on this risky adventure? This thinking was soon to change, at least in some circles.

The first officially sanctioned attempt to change this thinking took place in March 1962 when Charles P. Sonett, of the NASA Ames Research Center in California, was asked to convene a small group of scientists to recommend a list of experiments to be undertaken once the astronauts landed on the Moon. This meeting, requested by NASA's Office of Manned Space Flight, was held in conjunction with a National Academy of Sciences Space Science Board Summer Study taking place at Iowa State University in Ames so that the Academy's participants could review and comment on the recommendations Sonett's team would make. The Sonett Report, submitted to NASA management in July 1962, became the foundation for all subsequent lunar science studies and recommendations. Circulated in draft form at NASA and other organizations throughout the rest of 1962 and most of 1963, the report elicited both support and criticism. It is at this point in the evolution of Apollo science, with a short digression to set the stage, that I became involved, and here I take up the story.

Each chapter is written as a somewhat complete account of its subject. The chronology for a given chapter is correct as events unfolded, but there is some overlap in time as we move from one chapter to the next. I hope this will not be confusing but will provide a better perspective on how the individual pieces of the lunar science puzzle came together. I have also attempted to explain the roles of the key contractors and give credit to some who worked with us from the very beginning as we struggled to define and build the many experiments and supporting equipment that eventually made up the Apollo science payloads. I believe that most accounts of the Apollo program fail to give enough recognition to the many contractors who were essential contributors to the project's success.

One of the major players in this story was the late Eugene M. Shoemaker. Gene was involved in almost every aspect of Apollo science and had graciously agreed to review this manuscript when it was ready. I was greatly anticipating the comments and critique of this friend and colleague, hoping he could refresh my memory and suggest additions or changes for accuracy. But before I could send him an early manuscript, Gene died tragically in an auto accident on July 18, 1997, while studying impact craters in Australia. He will be fondly remembered and greatly missed. Not only was he an outstanding scientist who shaped our thinking on many subjects, including how we should explore the Moon, he was also a brilliant teacher whose greatest legacy, perhaps, will be the many young (and old) scientists and engineers who will follow in his footsteps and lead us back to the Moon and beyond—to Mars and the far reaches of our solar system.

Taking Science to the Moon

From the Jungle
to Washington

In February 1962 John Glenn was at Cape Canaveral preparing for his attempt to become the first American to orbit the Earth during the Mercury program. I was working for the Mobil Oil Corporation as an exploration geologist supervising a small field party in the rain forest of northern Colombia. Even in this remote area I could pick up Armed Forces Radio and the Voice of America on my battery-operated Zenith Transoceanic radio and stay up to date on the major events of the day. We had been closely following the launches of the newly formed National Aeronautics and Space Administration, and along with everyone back in the United States, we were disappointed at the failures and delays as we tried to catch up with the Soviet Union's aggressive space program.

After each of the several launch delays for Glenn's flight, NASA would project a new liftoff time, and based on these projections we would try to complete our daily fieldwork and get back to camp to hear the launch broadcast. Far from home, with our immediate world bounded by a small rain forest camp and how far we could ride each day on the back of a mule, it was easy to become absorbed in the drama at Cape Canaveral. One day, during one of the several holds before Glenn's launch, the announcer filled some airtime by interviewing someone from NASA's Public Affairs Office. During the interview Project Apollo was discussed (what little was known of it at the time), and it was mentioned that for the Moon landings NASA would need to hire geologists to help plan the missions. He gave an address where those interested could apply. My curiosity was piqued. I copied down the address, pulled out the rusty typewriter we used to write our monthly reports, and composed a letter to

NASA. I explained that I was not only a geologist but a former navy jet pilot and said I thought I would fit right in with NASA and all the astronauts.

Eventually John Glenn was launched successfully. When I next went to Bogotá I mailed my letter, convinced that NASA could not turn down such outstanding qualifications. In my naïveté I thought I might even have a chance to become an astronaut. Who had a better combination of experience to go to the Moon, I reasoned, than a geologist–jet pilot, especially one accustomed to working in strange places under difficult conditions (coexisting with army ants, vampire bats, and jaguars)? With some modesty, my letter implied this interest. It was several months before I had a reply from NASA—a polite letter thanking me for my interest. To be considered, I must fill out the enclosed forms and submit my application to the Goddard Space Flight Center in Greenbelt, Maryland. I did so, and the wait began—with some anticipation, given NASA's encouraging reply.

With the start of the rainy season I was back in Bogotá when another envelope arrived telling me I had qualified as a GS-13, aerospace technologist–lunar and planetary studies, and that my application was being circulated within NASA to determine if a position was available. I wasn't sure what an aerospace technologist was, but it sounded impressive. I had visions of being asked to do exciting things at this new agency with the improbable task of sending men to the Moon. Then began a longer wait. In December I received another letter saying that no positions were open but that they would keep my application on record in case one turned up. Rejection! That didn't fit in with my plans, and I resolved to pursue my quest the next time I was in the United States.

My next leave came in June 1963, and I decided to go to Washington to talk directly to someone at NASA. I bought an aerospace trade journal listing the latest NASA organization, complete with names. In it I found an office at NASA headquarters that sounded as if my background and interests would fit—Lunar and Planetary Programs in the Office of Space Science, headed by Urner Liddell. From my family's home in New Jersey I drove to Washington and, without an appointment, went to Liddell's office. He was traveling that day, but his deputy, Richard Allenby, was in. This was great good fortune, since Liddell turned out to be a rather formal bureaucrat who probably would not have seen me without an appointment. Dick Allenby was just the opposite and agreed to interview me. After briefly introducing myself, I learned he was an old oil field hand (geophysicist) who had worked in Colombia just a few years earlier, and we had several friends in common. We hit it off at once, marking the beginning of a long professional and personal relationship. Dick liked my background but had no openings. He then set up a meeting with navy captain Lee Scherer (another former pilot), who had just been hired to manage the Lunar Orbiter program (satellites that would orbit the Moon to photograph potential Apollo landing sites). He also was not hiring at the time, but he thought someone in the Office of Manned Space Flight needed a person with my experience. I was beginning to question my timing: lots was going on at NASA, with new offices being set up all over town, but just as the last NASA letter stated, no one had an opening. Lee, who would become my boss six years later, set up a meeting with another military man newly detailed to NASA, Maj. Thomas C. Evans, U.S. Army Corps of Engineers.

Tom Evans was an impressive officer, later to become a congressman from Iowa. Tom had been the officer in charge of establishing Camp Century in Greenland, the first successful adaptation of nuclear power for a military ground base. His background was ideal for his job at NASA—designing a future lunar base. After Lee Scherer's introduction got me in the door, he spent the next hour or so telling me about his new office's responsibility—planning a lunar program to follow a successful Apollo program. He was enthusiastic and brimming with ideas, the kind of leader everyone looks forward to working for. Best of all, he thought I could help the team he was putting together. Since it was getting late in the day, Tom asked me to return the next morning to talk to his deputy, Capt. Edward P. Andrews, U.S. Army, and determine how we could proceed.

My discussion with Ed Andrews went well, and since I had already received a civil service job rating, he proposed starting the paperwork to hire me. Two days in Washington and I was being offered a job as a lunar aerospace technologist at what I considered the most exciting place in town! It would mean a pay cut from my Mobil salary (I would receive the princely sum of $11,150 a year), but I couldn't pass up the opportunity. Ed took my paperwork and told me he would call me in Colombia when everything was final; he didn't see any reason the position would not be approved and said I should plan on moving my family to Washington.

Returning to Colombia in July, I took Ed at his word and began to close out my work. My supervisor knew about my plans, of course, since I had listed him as a reference. My coworkers all thought I was crazy to take on such a job; most thought trying to get a man to the Moon was quixotic at best and probably impossible. Planning what to do after we landed on the Moon was real science fiction. I thought they were all being short-sighted and that they would be missing out on the beginning of a real adventure. In August I got the phone call I was waiting for. Ed Andrews said all was in order and they were waiting for me to arrive. With a smug smile I filed away my NASA correspondence, including the rejection letter, and at the end of August my family and I left Colombia to begin a new calling—one that never lost its thrill and satisfaction over the next ten eventful years.

And so I began my career at NASA; a GS-13 aerospace technologist in the Office of Manned Space Flight, Manned Lunar Missions Studies. When I arrived in Washington, NASA offices were spread all over town awaiting the construction of a new government building dedicated to NASA, in southwest Washington. In September 1963 our offices were at 1815 H Street NW, just a few blocks from the White House. We shared the building with other organizations and other NASA offices, including program offices for manned planetary missions, systems engineering, launch vehicle studies, and other advanced studies.

I was assigned an office with another recent hire, Thomas Albert, a mechanical and nuclear engineer who was determining how to modify the planned Apollo systems to enable longer staytimes and lunar base missions. Since I came from a work environment where we primarily wrote reports based on work we had accomplished in the field or laboratory, Tom really impressed me. He would spend hours on the phone talking to NASA and private company engineers, taking a few notes and going on to his next call, all the while speaking a language I didn't understand, in which every third word seemed to be an acronym. I thought I'd never understand NASA-speak, in which acronyms were the order of the day. It was annoying at first, but soon I started to catch on and quickly moved to the next level where I invented my own program acronyms. This new skill brought a real sense of control. I am convinced that NASA could not have functioned without these shortcuts, and it became an unspecified requirement that new programs come up with catchy acronyms, most pronounced like real words, that would appeal to the ears and eyes of management, Congress, and the media. (You'll soon become accustomed to them as well and will have less need to consult the list of abbreviations in the front of the book.)

Our office at this time consisted of eight engineers with diverse backgrounds plus two secretaries. Except for Tom and Ed, we all shared the services of one secretary. Two or three engineers occupied each office space: new arrivals were assigned interior offices; offices with windows were for senior staff. Accommodations were spartan, but there were few complaints since we would soon be moving to a new building. There was one empty desk in the office I shared with Tom; it had been occupied part time by Eugene Shoemaker, detailed from the United States Geological Survey (USGS), who was on his way to Flagstaff, Arizona, to start a new USGS office. I missed meeting him by a few days, but our paths would soon cross, and we would work closely together until the end of Apollo.

My first days at NASA involved the usual getting acquainted. Although during my navy service I had been a part of another government bureaucracy, NASA functioned quite differently. Owing in part to Tom Evans's style and NASA's being a new agency with an unprecedented mission, multitudinous rules and procedures had not yet been instituted, and the staff was given great freedom of action. Since for the past six years I had usually made my own daily schedule, this was an ideal situation for me. With Ed Andrews's guidance I immediately began to define my role and make the contacts at NASA and in the scientific community that would make my job easier.

I soon learned that Gene Shoemaker had come to NASA to help bridge the wide gap between the science side of NASA, represented by the Office of Space Science (OSS), where I had made my first NASA contact, and the Office of Manned Space Flight (OMSF). Major differences had surfaced between OSS and OMSF over how to apportion NASA's overall budget. The debate on how to accomplish science on Apollo still lay ahead. OMSF was already receiving the major portion of NASA's budget, and OSS staff, as well as scientists outside NASA who looked to OSS to fund their pet projects, were constantly fighting to persuade top management to change NASA's funding priorities. These efforts were led by such luminaries as James Van Allen, who had made one of the first space-based science discoveries—the radiation belts surrounding Earth that were later named after him. The complaints were reinforced by the National Academy of Sciences and its Space Science Board, which provided advice to Homer Newell, the OSS administrator. I was told that Shoemaker, during his brief stay at NASA, had begun to reduce some of the distrust that had developed but had only scratched the surface. Apparently it would take more than his talents to resolve these differences. Despite many compromises and much cooperation, forty years later this power struggle still rages inside and outside NASA.

Into this controversial arena I ventured and, with Tom Evans's blessing, was given an unofficial second hat to work with both OSS and OMSF on matters dealing with lunar exploration. When Shoemaker left, Verne C. Fryklund, who had been working on Newell's staff, took his place. Fryklund was definitely from the old school. Gruff, with a bushy mustache and a half-smoked but unlit cigar perpetually in his mouth, he usually looked professorial in a tweed jacket with leather elbow patches. Being detailed from USGS, he was given the title of acting director, Manned Space Sciences Division, Office of Space Science. His primary duty was the same as Shoemaker's—to be the go-between for the Office of Space Science and the Office of Manned Space Flight. During his shuttle diplomacy, he was to present the interests of the science community to NASA's manned space side, which was not viewed as friendly to science. Fryklund became my unofficial second boss. By Washington standards his title was not imposing, especially with the "acting" designation. His staff was appropriately small, consisting of several headquarters staffers and a number of detailees, including geologist Paul Lowman from the Goddard Space Flight Center (GSFC) and several others from the Jet Propulsion Laboratory (JPL). Thus he was receptive to having me join his office.

Fryklund, an experienced bureaucrat, approached his new job cautiously. The complicated politics were self-evident to someone with his background, and he was fully aware of the gulf between the two organizations. Until this time nothing had been officially decided about what science projects would be carried out on the Apollo missions. This became his first priority. Shuttling back and forth between high-level meetings at OSS and OMSF, Fryklund relied on a draft report on the scientific aspects of the Apollo program (commonly referred to as the Sonett Report after its chairman, Charles P. Sonett of the NASA Ames Research Center).[1] It served as his guide and point of departure to lend weight to his arguments on what needed to be done for Apollo science.

Sonett's ad hoc working group had convened at Iowa State University in the spring of 1962 at the request of the Office of Manned Space Flight to recommend what scientific activities should be included on the Apollo missions. The group had twenty members and consultants with diverse scientific backgrounds, including strong representation from USGS led by Gene Shoemaker.

Paul Lowman served on the geophysics (solid body) subgroup and also helped compile the final report, while Fryklund worked with the geology and geo-chemistry subgroup during their meetings.

William Lee, assistant director for human factors in the Office of Manned Space Flight, provided guidelines at the start of the working group's delibera-tions. These guidelines defined the parameters within which the working group would operate. They were relatively short and simple (two and a half pages), since at that time little was known about the constraints the astronauts would be operating under and since all the Apollo hardware was in an early design phase.

The working group was asked to consider experiments and tasks that could be accomplished on the Moon in periods of one hour, eight hours, twenty-four hours, and seven days. Because NASA still was not sure what the flight profiles would be, no guidance was given for any operations on the way to the Moon or in lunar orbit. Choosing landing site(s) was also not part of the working group's charter, although its recommendations could influence site selection. Advice on power and communication capabilities for transmitting scientific data was very general, and the committee members were told that this should not restrict them. They were to plan for more than one but fewer than ten missions with the possibility of carrying one hundred to two hundred pounds of scientific pay-load. Life-support supplies would limit the crew's operations to a radius of approximately half a mile. They were cautioned that the astronauts' space suits might hinder their ability to perform "precise manipulations." And finally, they were told that it might be possible to include a "professional scientist" in the crew, but that this would "significantly complicate our selection and training program, and [such a recommendation] should not be made unnecessarily."

Today, reading between the lines and looking at the numbers the committee was given to work with, it seems clear that these guidelines sent a message to the members that scientific ventures during the Apollo missions might be tolerated but that they should not have high expectations. This message was repeated in the years ahead, much to the dismay of the scientific community.

Despite the restrictions, the draft report contained wide-ranging recom-mendations that included geological and geophysical experiments to be done on the Moon as well as experiments in surface physics, atmospheric measure-ments, and particles and fields. Bill Lee's guidelines were to some degree ig-nored; the assembled scientists could not resist telling NASA what needed to be done. What they recommended could not be carried out with only one to two hundred pounds of payload, and they described geology traverses up to fifty miles from the landing site. They also detailed sample collection, including drill or punch core samples, and potential landing sites were suggested by Shoemaker and by Richard E. Eggleton of USGS and Duane W. Dugan of the Ames Research Center. The report went so far as to describe what type of astronaut should be on the flights and the criteria for finding such recruits.

Since the report had been requested by OMSF and not by the science side of NASA, its recommendations carried some weight in OMSF offices. The draft had been circulated to participants at the National Academy of Sciences 1962 Iowa Summer Study, who had met at the same time as Sonett's working group.[2] Thus the Sonett Report would include the endorsement of the other side of NASA's house (the scientists) when it was officially released. Although the Iowa Summer Study group agreed with the general conclusions of the Sonett Report, it recommended that the scope of the proposed investigations be more restricted than those spelled out in the report, a rather surprising recommendation in light of later criticisms from the scientific community.

Based on these recommendations, and with his bosses in both OSS (Homer Newell) and OMSF (Joseph Shea) concurring, in early October 1963, one month after my arrival, Fryklund sent a memo to Robert R. Gilruth, director of the Manned Spacecraft Center (MSC) in Houston, containing the first official scientific guidelines for Project Apollo. As is the nature of guidelines, they established a broad framework for planning, but they provided no specifics on how long the astronauts would be on the Moon or how much payload weight should be allocated for science. These numbers were to come later. The eight guidelines included a listing of three functional scientific activities in decreasing order of importance: "a. Comprehensive observation of lunar phenomena; b. Collection of representative samples; and c. Emplacement of monitoring equipment."[3] Assigning sample collection a number two priority is interesting since, as we will see, in later planning it became the astronauts' first task once they were on the lunar surface. Back in Washington we began trying to flesh out the guidelines by reading between the lines of the Sonett Report and translating the recommendations to some hard numbers.

From the information we could collect, it was evident that the range of measurements and activities the Sonett committee had listed, even if reduced to follow the National Academy of Science's recommendations, would require a science payload far exceeding the target of one to two hundred pounds. One month before Fryklund issued the guidelines, and unknown to headquarters, MSC jumped the gun and hired a contractor, Texas Instruments, to spell out Apollo experiments and measurements to be made on the lunar surface based on MSC guidelines. The report, when it was eventually issued in 1964, was dismissed as amateurish by headquarters and by members of the scientific community who had begun to focus on Apollo science. This difference of perspective signaled a clash between headquarters and the small MSC science staff over who would define Apollo science.

Adding to this mix of ideas on what science to carry out on the Moon, in early 1963 Bellcomm engineers had provided some analyses of potential Apollo and post-Apollo scientific operations. Bellcomm had been created in March 1962 by AT&T at the request of NASA administrator James Webb to provide technical support to NASA headquarters. By the time I arrived Bellcomm had grown to over 150 engineers and support staff and had already run afoul of MSC engineers, who accused the company of being a meddling tool of head-quarters—some at MSC went so far as to call the staff headquarters spies. MSC tried to exclude them from some meetings by keeping the schedules quiet so that when the meetings were announced it would be too late for the Bellcom-mers to make the trip from Washington to Houston. Another aspect of the visits that MSC found annoying was that Bellcomm required trip reports, so everyone who read them knew about what went on and about any disagree-ments with MSC's proposals. Disagreements were frequent, and the second-guessing by Bellcomm continued throughout the program, often leading to positive changes, especially concerning the science payload. Eventually a small group of Bellcomm scientists and engineers were assigned to support Evans's office, and they became important adjuncts to our small staff. Their support and numbers grew as Apollo science evolved.

At the end of January 1963 two Bellcomm staffers, Cabel A. Pearse and Harley W. Radin, presented a study examining the scientific advantages of having an unmanned logistic system deliver a fifteen-hundred-pound payload to the lunar surface. They concluded that the best use of such a system would be to provide "a fixed scientific laboratory equipped with a wide variety of sci-entific instrumentation."[4] Two months later, under the leadership of Brian Howard, one of England's "brain drain" expatriates, with Robert F. Fudali, Cabel A. Pearse, and Thomas Powers, Bellcomm issued a second report, *The*

Scientific Exploitation of the Moon.[5] It provided a preliminary analysis of the type of science that might be conducted utilizing Apollo hardware to deliver a logistics payload of seven to ten thousand pounds to the lunar surface, the payload sizes being studied by Evans's office. Although the second report does not cite the draft Sonett Report by name, the authors were surely aware of its existence because they include most of the experiments it described and it is cited in the January report. In addition, they recommended carrying out a variety of other operations and experiments including the use of roving vehicles and deep drilling. To my knowledge the Bellcomm reports and *Lunar Logistic System,* a ten-volume report issued by the Marshall Space Flight Center (MSFC) at the same time as the Bellcomm report, represent the first attempts to document the feasibility of using Apollo hardware for extended exploration on the lunar surface.[6] These reports were my first exposure to such thinking and were among the early references on my NASA office bookshelves.

In late October 1963, returning from one of these frequent meetings, Fryklund rushed into the office we shared and announced, "They've just agreed; we have 250 pounds for science!" "They" being NASA Manned Space Flight senior management. Having been on the job only a few weeks and a latecomer to what had been a major struggle, I showed only muted enthusiasm. Based on my limited experience and initial looks at what a good science payload like that recommended in the Sonett Report would weigh, 250 pounds seemed a minor victory. A thousand pounds or more would have been better. But a victory it really was, certainly better than the one to two hundred pounds given to the Sonett working group. Once our foot was in the door, we quickly capitalized on the opportunity to define a complete payload based on this "official" number.

Other major changes had also been taking place in NASA. Headquarters was swiftly evolving. New organizations were being created almost weekly, and the staff was expanding rapidly. During 1963, the year I came, NASA headquarters almost doubled in size. With all these changes the headquarters phone directory was always out of date, and addenda were published every month. Brainerd Holmes, who until September had been in charge of manned space flight operations as director of the Office of Manned Space Flight, resigned and was replaced by George Mueller from Space Technology Laboratories. Mueller was given the new title of associate administrator, Office of Manned Space Flight, a third tier of top management just below administrator James Webb and his deputy, Hugh Dryden and associate administrator Robert Seamans. Homer

Newell was elevated at the same time to a similar position with the title associate administrator, Office of Space Science and Applications (OSSA). With his appointment Mueller introduced a different management style to Manned Space Flight, one that would have a profound effect on Project Apollo's future.

Toward the end of the year our office was merged with several others, and the new organization was called Advanced Manned Missions Programs. Edward Z. "E. Z." Gray was hired from the Boeing Company to be our leader, and we soon moved to our new offices at 600 Independence Avenue SW. In January 1964 Maj. Gen. Samuel C. Phillips was detailed from the Air Force Ballistic Systems Division to become Mueller's deputy director for the Apollo program. Later in the year his title was upgraded to director.

In the wave of reorganization, Fryklund's tenure as acting director was short lived. Homer Newell, in agreement with Mueller, formally established the Office of Manned Space Science, reporting to both his office and Mueller's. Willis Foster was brought in from the Department of Defense as the new full-time director, and Fryklund became Foster's chief of lunar and planetary sciences. After some eight months working for Foster, he transferred back to Newell's staff, and a short time later he returned to USGS to work in its military geology branch. Foster's office, starting with an original staff of eight, grew rapidly (and now included Peter Badgley, my former thesis adviser at the Colorado School of Mines). Dick Allenby was transferred from the OSSA Lunar and Planetary Programs Office to become Foster's deputy. Anthony Calio was brought in from the newly formed Electronics Research Center in Cambridge, Massachusetts, to provide some engineering muscle, and along with Jacob "Jack" Trombka he began to coordinate the planning for scientific instrumentation. Edward Chao, another USGS detailee, became the office expert on how to handle the anticipated scientific treasure—the samples collected. Edward M. Davin, an acquaintance of Allenby's, was hired from Esso Research (now Exxon) in Houston in the summer of 1964 to join Allenby as the resident geophysicists, representing a scientific discipline that would increase in importance as the Apollo experiments were selected.

Will Foster now became my unofficial second boss, and I continued to work on developing the science payloads for Apollo flights as well as later undertakings. How we accomplished this for Apollo, and eventually went far beyond the initial 250-pound allocation, follows in the next chapters. But first, from a scientific perspective, why fight to get a science payload on Apollo in the first place?

Early Theories and Questions
about the Moon

If you have binoculars of ten power or even less, you can go out in your backyard on any clear night when the Moon is up—best perhaps at a quarter-moon phase, not a full moon—and become a lunar scientist. Brace yourself against a solid support so your hands are steady and focus on the line that separates the illuminated part of the Moon from the dark portion. Near this line the Sun casts the longest shadows, and you can see the greatest topographic detail. The technical term for this line is the lunar terminator, but you needn't know this to start your studies. Your ten-power binoculars are about half as powerful as the telescope constructed by Galileo Galilei, who early in the seventeenth century first began to study the Moon with more than the naked eye.

What will you see? Depending on where the line between the bright and dark portions falls on the particular night, you will probably see, just as Galileo did in 1609—to his amazement—some large and small circular craters, perhaps some mountains, and some apparently smooth areas that are known as maria, or seas. In 1963, some 350 years after Galileo made his first observations, the craters were the most controversial of all lunar features, sparking the most heated debates. What was their origin? Were they the remains of volcanoes? Were they caused by impacts like those that left similar craters on Earth? Were they the result of some combination of processes or the product of unknown forces? The lunar maria were also controversial; they were generally interpreted as lava flows. But how were they formed, and how did they spread over such a vast area? How were the mountains formed? Their very existence provoked debates about the internal structure of the Moon and its evolution.

The major, fundamental lunar questions being debated by planetary scientists when the Apollo program began can be quickly summarized: How old is the Moon, how was it formed, and what is its composition? Finding the answers was the driving force behind the desire to carry out a host of experiments on the Apollo missions. And a large science payload would be needed to resolve these difficult questions. The answers to some of them would come in part from samples collected on the Apollo landings, and in turn the samples would tell us a lot about the origin of the craters. If the Apollo missions landed at interesting points on the Moon and included various geophysical experiments along with geologic traverses, these mysteries might be resolved. From the answers we anticipated understanding Earth better, especially its early history. When I joined NASA in 1963 my knowledge of the Moon and of the ongoing debates was close to zero. I quickly resolved to fill this void and began to study the literature.

As soon as I returned to the United States from Colombia, I went to the local library and bookstores to find books to increase my meager knowledge. To my surprise, there were very few. And in recalling my undergraduate and graduate studies in the earth sciences, I could not remember that any attention had been paid to the Moon or the Earth-Moon system. The first book I bought was *The Measure of the Moon*, by Ralph B. Baldwin.[1] It turned out to be a fortuitous choice. Not only had Baldwin done a comprehensive survey of the literature (the specialized literature was much more extensive than that found in general bookstores), he had organized the existing knowledge and theories and presented them in a readable fashion. His opening sentence was prophetic: "Every investigation of the Moon raises more problems than it solves." During the next five or six years I would find myself immersed in these problems and dealing daily with the various protagonists cited in the research. I later learned that I was in good company by being impressed by Baldwin's work; Harold Urey, a Nobel laureate in chemistry, had become fascinated by the Moon's many mysteries after reading Baldwin's earlier book, *The Face of the Moon*, and had put forth his own theories on how the Moon formed.

My first impression that there was little source material quickly changed. Baldwin's references were extensive, too many—in light of my new duties—for more than a cursory review. I settled on purchasing a few texts to read in their entirety and keep available as a small reference library. In addition to Baldwin I read *The Moon*, by Zdenak Kopal and Zdenka Mikhalov; *Structure of the Moon's Surface* by Gilbert Fielder; Harold Urey's *The Planets* and several of his articles and reports; Gerard P. Kuiper's "On the Origin of Lunar Surface Features"; and an article by my old mines professor L. W. LeRoy, "Lunar Features and Lunar Problems."[2]

Perhaps most interesting of all, I discovered that most of the leading figures in lunar and planetary science, including Urey, Kuiper, Fielder, Kopal, and Baldwin, were active and accessible. In addition, some of the younger lions, such as Shoemaker, Frank Press at Lamont-Doherty, Jack Green at North American Aviation, John O'Keefe at the Goddard Space Flight Center, and Carl Sagan of Cornell University, were already involved in NASA programs.

The origin and age of the Moon had intrigued astronomers and Earth scientists for many centuries, with theories proposed based on a minimum of hard data. By the early 1960s existing theories had become more sophisticated, supported by ever increasing observational data and, soon, by returns from several of NASA's unmanned programs. Three theories on the Moon's origin held sway: (1) the Moon and the Earth had formed more or less simultaneously from the same primordial cloud of debris surrounding the Sun; (2) the Moon had been separated from the Earth either through tidal movements or by the impact of another body (some would split this into two theories); and (3) the Moon had formed elsewhere in the solar system, and in its orbit around the Sun it had been captured by the Earth's gravitational field in an early close encounter. Based on the information then available, each of these theories could be supported or argued against depending on one's point of view and which data one considered most critical. The date when any of these events took place was also conjectural, but it was generally believed that the Moon had become Earth's companion early in the formation of the solar system, some 4.5 to 5 billion years ago.

Certain information was well documented. The Moon's physical dimensions and mass, its distance from Earth, and several other properties were known rather precisely. Unlike Earth's, the Moon's magnetic field, if any, was thought to be weak; its mass and volume translated to a body less dense than Earth, probably without an iron core or at best with a very small core. It had no discernible atmosphere. We knew that the Moon was locked into a slowly expanding orbit that allowed only one side to face Earth. The Moon's farside or back side (not "dark side" as so many ill-informed writers call it, since it is lit by the Sun in the same manner as the side facing Earth) was a total mystery; was it the same as what we could see or very different? This lack of information had made the Moon's farside the playground of science fiction writers for many years. One could imagine all sorts of strange things back there, including alien colonies.[3]

Probably the most contentious issue was the origin of lunar craters. Were they formed by some internal process like volcanism or by the impacts of small to large bodies like meteorites? The literature was full of this particular controversy, and the debate—at times vitriolic—went on at all lunar symposia. Each side had its champions, although it appeared that the "impactors" were beginning to win the day. Any of the three lunar origin theories could accommodate either an impact or a volcanic explanation, but the subsequent history or postorigin modification of the Moon's surface would be entirely different depending on which crater theory proved correct. If the craters were volcanic, then the Moon's interior had been molten after its formation and we could expect to find many Earthlike conditions. If the craters were caused by impacts, then the Moon's evolution might have been very different from Earth's, even though most students believed that impacts were common in Earth's early history. Complicating this debate, we could observe other features on the Moon such as sinuous, riverlike rills and odd-shaped depressions that did not conform to the contours expected of impact craters. What was the Moon trying to tell us? Had there once been water on the Moon? Had a combination of processes taken place? Were they still taking place?

A primary scientific justification for studying the Moon, with either manned or unmanned spacecraft, was to help us unravel Earth's early history. A new term had been coined for such study, "comparative planetology," and we used it frequently in our briefings both inside and outside NASA. Comparative planetology means studying the planets by comparing what can be observed or measured on one with similar characteristics on another; through this back-and-forth association we would increase our overall understanding of all the planets. We believed that applying this technique to the Earth-Moon system would be especially fruitful. In all the solar system, our Moon is the largest relative to the size of the planet it orbits—in essence we are a two-planet system. By studying the Moon we believed we would learn much about Earth. When the Apollo project began many basic questions concerning our home planet were unanswered, and many were similar to those we were asking about the Moon. How was it formed, and how had it changed during its early evolution? What is the thick zone just beneath Earth's crust—the mantle—made of?

How does the mantle influence or produce the energy that moves large sections of Earth's surface?[4]

Earth's surface is a dynamic place. Mountains rise and are eroded away, sea basins and lakes fill and dry up, and continents move vast distances, a process called "continental drift." The record can be deciphered by earth scientists in the rocks of Earth's crust. But our understanding becomes sketchier and more uncertain as we go back in time toward Earth's earliest history. That part is obscured, hidden, or even destroyed by the very processes mentioned above. The oldest Earth rocks that have been positively dated, from northern Canada, are approximately 4 billion years old. The oldest piece of the solar system dated thus far is the Allende meteorite, calculated to be almost 4.6 billion years old, supporting the earlier theories that the solar system might be 5 billion years old. These dates, however, leave a gap of almost a billion years from the oldest dated Earth rocks to the solar system's birth. This billion-year gap continues to be an enticing field for speculation and investigation.

Returning now to the three theories of lunar origin: What were their implications for Apollo? Could we expect to shed light on these riddles or perhaps even solve them? If either of the first two was correct—if the Earth and the Moon formed simultaneously and close together or if the Moon broke off from Earth, then one would expect the rock types or minerals we would find on the Moon to be similar to those on Earth. If the third theory was correct, that the Moon formed somewhere else in the solar system and was later captured by Earth, then we might find different rock types and minerals on the Moon, perhaps similar to some of the more exotic meteorites that have been recovered at various places on Earth. Regardless of the ultimate answer, we were confident we would be able to date the rocks and get a handle on a pressing question: When was the Moon formed? Some believed the Moon's surface was ancient, that all the features we observed had formed early in its history and had changed little since then. Confirming this would be exciting; the Moon, as many were fond of saying, could act as a Rosetta Stone in deciphering the birth of the Earth and the solar system!

Harold Urey at the University of California, San Diego, was a strong proponent of the third theory. He believed the Moon had been formed through the accretion of planetesimals (large pieces of the primordial cloud from which the Sun and eventually the whole solar system evolved) and that this happened some 4.5 billion years ago. If true, it was an ancient and unchanged body and worthy of careful study. The Moon has an irregular shape (it is not a perfect spheroid but has slight polar flattening and an Earth-facing equatorial bulge), and it wobbles on its axis. Urey argued that the Moon had never been completely molten or these irregularities would not have survived. According to his calculations, the Moon had formed as a somewhat cold body—those who said the maria were lava flows erupting from a molten interior were wrong. The maria, he believed, were the result of large-scale melting caused by the impact of large bodies, such as the one that had formed Mare Imbrium, and the maria material might have been the melted remains of carbonaceous chondrites, an unusual type of meteorite occasionally found on Earth. Urey was looking forward with great anticipation to obtaining lunar samples, especially from the maria (they should not be Earthlike lava), to prove his theory.

Urey's reputation as a Nobel laureate was important in legitimizing our lunar studies. When he spoke, everyone listened. Although he had many differences with other lunar scholars, sometimes he agreed with them. He agreed, for instance, that most craters were certainly of impact origin and that much of the lunar topography was shaped by ejecta from the impacts. He did not think there had been much volcanism on the Moon, but he accepted the observations of some volcanolike features. In a letter to Jay Holmes at NASA headquarters in January 1964 Urey said: "I am sure that only the most experienced hard rock geologist could possibly do anything about the subject satisfactorily. I urge strongly that all astronauts be well trained hard rock geologists. The Apollo project is being severely criticized by outstanding people, and I believe that if we do not at least [do] the very best that we can to solve important scientific problems that this criticism may well swell to a very great chorus."[5] Urey's suggestion on astronaut training was one of the first shots in a long campaign that led to the scientist-astronaut selections discussed in later chapters. Regardless of his opinions, his presence at any lunar symposium guaranteed vigorous debate and lots of publicity, a commodity we eagerly sought as we struggled to make NASA management recognize how important the Moon would be in resolving issues of such magnitude.

Another vigorous debater was Thomas Gold, a professor at Cornell University who had made his early reputation in astronomy. In recent years he had focused on problems related to the Moon. Tommy Gold was to prove a thorn in our sides with his strange theories, seldom supported by anyone else in the scientific community. His most controversial one, first proposed in 1955, was that the lunar surface was covered by a layer of fine particles eroded from the lunar highlands, perhaps several kilometers thick, that could move across the lunar surface and fill in depressions.[6]

He sought to prove this contention with photographs showing that most lunar features had a smooth appearance and many craters seemed to be filled rather uniformly with some material. He generally discounted the idea that this fill might have been molten material like lava or ejecta from impacts. Radar studies of the Moon tended to support his thesis that the uppermost soil layer was fine grained and of low density, but how thick this layer might be and what area it covered could not be resolved from the radar data.[7] Other interpretations were also possible.

The character of the lunar soil, especially its topmost layer, was of course a great concern, since it would directly affect the design of the lunar module (LM) and the astronauts' ability to land and move around on the surface. Not much was known about how soils and fine-grained material would behave in the high vacuum found on the Moon. Several government and private laboratories had done experiments to examine this question. Bruce Hapke at Cornell University, for example, had shown that fine particles deposited in a vacuum tended to stick together loosely, forming what he called "fairy castle" structures, or soils with low bearing strength.[8] This could be seen as substantiating Gold's contention of a low density lunar surface.

Before the return of pictures from Ranger, and later the Surveyor and Lunar Orbiter missions, photographs of the Moon had come from telescopic images, with a resolution of at best a thousand feet. Under such low resolution, every feature on the Moon appeared somewhat smooth. This problem did not deter Gold. Even after we received the higher resolution Ranger, Surveyor, and Lunar Orbiter photos, he continued to predict that when the lunar module attempted to land it would sink out of sight in his electrostatically levitated dust. At this early stage such predictions alarmed NASA's engineers, for it was difficult to prove him wrong.

Fortunately questions of this type—though not so outrageous—had been anticipated, and the Surveyor spacecraft were designed to answer them. Surveyor did prove Gold wrong, which he accepted grudgingly, continuing to maintain that some areas of the Moon were covered with fluffy dust. He clearly enjoyed being the center of controversy, and after Surveyor's deflator he came up with another whopper: the lunar dust would be pyrophoric. When the astronauts landed and opened their LM hatch, the oxygen released from the cabin would combine with the soil and cause an explosion. His reasoning was that the lunar surface, exposed for eons to the bombardment of the solar wind, had become oxygen deficient and would undergo an explosive oxidation when exposed to the LM atmosphere. This prediction also worried the engineers, and it would not be possible to prove or disprove it with any projects in the pipeline before the actual landing.

The school of volcanic crater supporters started strongly and slowly declined in influence as more and more observational and experimental data became available. But in 1963 and 1964 they still made a good case for their views. The leaders of this school were Gerard Kuiper, at that time director of the Lunar and Planetary Laboratory in Tucson, John O'Keefe at Goddard Space Flight Center (GSFC), and North American Aviation's Jack Green. Each of these advocates had a somewhat different interpretation of what was observed on the Moon. Both Kuiper and O'Keefe admitted that impacts had played a role in the Moon's evolution, but they still thought volcanism was the major explanation of its present surface formations. Kuiper had been an early student of the Moon. Ignoring Urey's counterarguments, he believed the original substance that came together to form the Moon contained enough radioactive material to eventually raise the interior temperature and melt the entire Moon. In his model this had occurred some 4.5 billion years ago, forming the maria and filling the larger craters, all subsequently modified by meteoroids.

Green, however, took a hard-line approach. Essentially all features on the Moon could be, and should be, explained by volcanic processes. Jack was a colorful figure, never taken aback by criticism, and a good debater. You could count on him to enliven any lunar symposium. His forte was showing side-by-side photographs of terrestrial and lunar features that looked almost identical. The terrestrial features, of course, were always volcanic in origin.

The impact school was led by Gene Shoemaker and his United States Geological Survey (USGS) followers. Gene had been influenced by an earlier and revered USGS chief geologist, Grove K. Gilbert, who in 1893 published a paper concluding that the Moon's craters were probably of impact origin.[9] Gene had carefully studied Meteor Crater in Arizona, just east of his new Flagstaff offices, as well as several other craters of known impact origin in other parts of the world. Robert Bryson, from NASA headquarters, had funded Gene to develop a detailed report of his findings that would combine his earlier studies and field observations at Meteor Crater. By 1964 Gene's studies had been completed for some time, but he had not finished the written report. This was a sore point with Bob because so little had been published on the geology and mechanics of impact craters, and Gene's work was intended to fill this void. He had published a short report on his work in 1963, but the full report was still in draft form.[10]

Bob, a former USGS geologist, had great insight into what it would take to convince the scientific community that important information could come from lunar studies. In addition to Gene's work, Bob funded some of the studies of Ed Chao at USGS, who in 1960 discovered coesite in the shocked debris from Meteor Crater, a type of silica that forms only under extremely high pressure. Before Chao's discovery, coesite had been made in the laboratory but had never been found in nature. This mineral is now a key fingerprint for identifying impact craters. Soon after this discovery Chao found stishovite, another form of high pressure silica, in rocks ejected from Meteor Crater—further confirmation that an impact of enormous energy had created the crater. Chao was later detailed to NASA as Apollo science work expanded, and we worked together under Will Foster. Bryson also funded the telescopic mapping of the Moon, initially through Robert Hackman at USGS. These maps laid the groundwork for all the subsequent lunar geological interpretations used during the Apollo landings and the planning that preceded them.

Despite the annoyance at NASA headquarters about the Meteor Crater report, Gene was a walking encyclopedia concerning what happens when a relatively small meteorite hits a solid object like Earth. (The iron meteorite named the Canyon Diablo that blasted the four-thousand-foot-diameter Meteor Crater probably weighed about seven thousand tons.) He extrapolated these results to the larger lunar craters that must have been formed by even larger bodies. He was joined in this knowledge by experimenters such as Donald Gault at the NASA Ames Research Center and others who had conducted small hypervelocity, laboratory-scale impact studies. In addition to making direct field observations on Earth, Gene and his staff, following Bob Hackman's lead, had spent considerable time mapping the Moon using several large telescopes. Applying standard terrestrial geological interpretations to these eyeball studies, they had become convinced that the Moon was pockmarked with impact craters. Shoemaker was sure that almost all lunar craters had been formed by this mechanism, not through volcanism.

In a trip report of a visit to Menlo Park in May 1963, Bob Fudali described his conversations with Henry Moore, Dick Eggelton, Donald Wilhelms, Harold Masursky, and Michael Carr of USGS.[11] After spending many hours drawing geological maps of the Moon based on telescopic observations, the USGS geologists believed that, despite the high density of impact craters, there was substantial evidence of volcanic activity on the Moon, somewhat at variance with Shoemaker's views. They also believed there was evidence that the maria were covered with extrusive igneous material, and they were convinced that tektites (rounded glassy bodies probably of meteoritic origin found at several places on Earth) originated on the Moon, thus supporting O'Keefe's theories. Because of the chemical composition of the tektites, this meant that at least some parts of the Moon were "granitic," which in turn meant that at some point in its evolution the Moon had undergone differentiation in the presence of water. One could then conclude that the Moon was at least somewhat like Earth.

In addition to these major theories and vigorous debates, several related questions had puzzled lunar scientists for many years. Answers were especially important to the new breed of comparative planetologists, for they hoped the answers would shed new light on similar questions about Earth's evolution.

During its early formation, Earth went through partial melting and differentiation. As the material that was to make up the bulk of Earth's mass accumulated, the heavier material sank to the center, forming a core. Each layer above the core was of decreasing density, and the lightest materials formed the crust. Although we do not completely understand these various deep materials that form the bulk of Earth's interior, we can infer and calculate what they are. Based on this knowledge, we have reconstructed the processes that formed them. As an example, we know that Earth's continents are relatively light material "floating" on denser underlying rock. We also know that through geologic time there has been a constant churning of the upper layers and that Earth's surface today looks very different than it did, say, 3 billion years ago. Although we say we know these things, they are really just theories based on observable field data and hypothetical calculations. It would be reassuring if we could find other examples of these processes or similar ones elsewhere in the solar system. What better place to look than the Moon, our closest neighbor?

Had the Moon undergone differentiation in its early history? Telescopes showed mountains on the Moon. They were generally lighter in color than the lowland maria and thus probably different in composition. Were the mountains less dense, as terrestrial mountains are less dense, on average, than Earth's crust and upper mantle? If you believed that tektites came from the Moon, differentiation was a given, with less dense material occurring at the surface. Did the Moon have a core? The tiny but measurable magnetic field (averaging five gammas and believed to be due primarily to the interaction of the Moon with the solar wind) and overall lower density seemed to negate a lunarwide field, but we had not been able to make close-up measurements. Perhaps there were weak, relict local magnetic fields that would be evidence of early core formation. Why did the nearside and farside of the Moon look different? This question became more important when we received Lunar Orbiter pictures of the Moon's farside with much higher resolution than those returned by *Lunik 3* and the full extent of these differences became known. Did Earth-Moon tidal effects account for these differences, or was it some other factor?

Whether water ever existed on the Moon was another important question. Because the Moon has no discernible atmosphere (it was estimated to be equivalent to Earth's atmosphere at altitudes above six hundred miles, appropriately an exosphere),[12] water probably would not be found on the lunar surface under any conditions, but it might still exist belowground. Some proposed that it might be found in permanently shadowed craters near the lunar poles. Urey in 1952 and Kenneth Watson, Bruce Murray, and Harrison Brown in 1961 provided an analytical basis for such predictions. The latter authors concluded, "In any event, local concentrations of ice on the moon would appear to be well within the realm of possibility. Unfortunately, if it exists, it will be found in shaded areas, and attempts to determine whether it is present must await the time when suitable instruments can be placed in those areas."[13] Some thirty-five years later the Clementine and Lunar Prospector missions seem to support their analysis, though it is probably safe to say the authors had not imagined that ice would be detected by instruments in lunar orbit; such a possibility was beyond their dreams in the early 1950s.

On Earth, water is needed to form granites, so if granites existed on the Moon, then water must have been present in its early history. If water could be found on the Moon it would greatly simplify our plans for post-Apollo manned exploration. Its presence in an easily recoverable form would reduce the potable water we would have to transport to the Moon, and water could be used as a source of oxygen for manned habitats. Far-out planners even envisioned making rocket fuel by separating the hydrogen and oxygen. The questions posed by present-day space planners or raised by the information gained from the Clementine and Lunar Prospector missions thus are not new but were on our minds thirty years earlier.

Would we find any evidence of life forms, however primitive, in the samples brought back to Earth? This outcome was considered unlikely but not impossible. For this reason the samples and the astronauts would be quarantined on their return lest they carry some deadly virus or pathogen to which we poor earthlings would have no immunity. Any evidence of life would be astounding and would require rethinking how life formed on Earth.

All the questions above, and their answers, were important both to NASA (especially my office) and to the scientific community in general. Our post-Apollo mission strategies were based on attempting to find answers, which in turn would help us plan our programs for Venus, Mars, and beyond, using the Moon as a staging point for these more difficult missions. And there was still the link to understanding Earth.[14]

All these theories, questions, and debates could be resolved by a relatively small suite of activities and experiments. The trick would be to design them so they could be carried on the missions and deployed by the astronauts. The astronauts would have to sample the rocks and soil at their landing sites over as large an area as possible and bring the samples back to Earth for analysis and reconstruction of their geological context. Also, to complete the picture they would need to carry certain geophysical instruments to collect data pertaining to the Moon's subsurface or other environmental conditions. In the introduction to his book, Baldwin had stated: "It is beyond hope that we shall ever have a complete and definitive answer to all lunar problems." Finally he had predicted: "Landing on the moon and analyzing its materials will help greatly but will raise more problems than are solved."[15] These predictions echoed concerns raised in his first chapter. We hoped that our plans for extensive manned lunar exploration would go a long way toward changing his mind on both of them.

After becoming reasonably familiar with the current state of knowledge about the Moon, I started making some personal observations. I got permission from Tom Evans to contract with the Astronomy Department at the University of Virginia for time on their large (twenty-six-inch refractor) telescope so some of us on the NASA headquarters staff could travel to Charlottesville and make our own close-up studies. Laurence Fredrick, director of the Leander McCormick Observatory, was a gracious host for those of us that took advantage of the opportunity. This telescope, almost a twin to the famous Naval Observatory telescope in Washington, D.C., where some of the first lunar studies had taken place in the nineteenth century, including those by Gilbert, was the one USGS used in 1961 to begin the detailed mapping of the Moon funded by Bob Bryson. Because this work had recently been transferred to the Lick Observatory in California and a new observatory near Flagstaff, observing time was available. The Virginia telescope was an ideal instrument for casual Moon viewing because with easily mastered techniques it provided a resolution of a few thousand feet for lunar surface features. Charlottesville was only a two-hour drive from Washington, so we could leave the office immediately after work, stop for a quick dinner, set up the telescope in plenty of time for a few hours of viewing, and still get home shortly after midnight.

A twenty-six-inch-refractor telescope is a very large piece of equipment. The telescope with its mount weighed some eight tons. A rotating dome with sliding doors covered the telescope, and housed within the dome were the electronics and motors that allowed one to point and track the telescope. Under Larry Fredrick's tutelage, I became adept at operating the instrument, and after a few nights' practice I was able to observe by myself. As one might expect, viewing was ideal on clear nights, and the winter months were best of all because cold, stable air reduces atmospheric disturbances. But even on exceptionally clear nights there was always a shimmering distortion caused by Earth's atmosphere, making it appear that heat waves were rising from the Moon and tending to obscure features under high magnification. I spent many a cold night studying the Moon's surface, following the terminator as it slowly moved across the face of the Moon revealing the surface detail. When the Sun's angle was correct I could compare my observations with the first USGS lunar maps of the Copernicus and Kepler regions to understand how this latest attempt to map the Moon geologically was carried out and why the USGS mappers were identifying certain types of surface features as discrete geological formations. The subtlety of most of these features was evident, and I came to appreciate how an earthbound geologist's imagination might become a dominant factor in drawing a geological map of the Moon with the enormous disadvantage of never having set foot on the surface.

Another compelling reason for spending time observing the Moon was the recent spate of reports by reputable astronomers about transient phenomena on the lunar surface. In 1958 a sensational announcement had been made by Soviet astronomer Nikolai Kozyrev, who claimed he had recorded spectra of a transient event on the Moon near the central peak of the crater Alphonsus. Other observers soon reported color changes and similar events at other lunar features, the most exciting being at the crater Aristarchus.

Excerpts from the report written by James Greenacre, employed at that time by the U.S. Air Force Lunar Mapping Program at Lowell Observatory near Flagstaff, Arizona, tell his exciting story of what he observed one night at Aristarchus.

> Early in the evening of October 29, 1963, Mr. Edward Barr and I had started our regular lunar observations. . . . When I started to observe at 1830 MST . . . I concentrated on the Cobra Head of Schroeter's Valley. . . . at 1850 MST I noticed a reddish-orange color over the dome-like structure on the southwest side of the Cobra Head. Almost simultaneously I observed a small spot of the same color on a hilltop across the valley. Within two minutes these colors had become quite brilliant and had considerable sparkle. I immediately called Mr. Barr to share this observation with me. His first impression of the color was a dark orange. No other color spots were noted until 1855 MST when I observed an elongated streaked pink color along the southwest rim of Aristarchus. . . . at approximately 1900 MST I noticed the spots of color at the Cobra Head and on the hill across the valley had changed to a light ruby red. . . . I had the impression that I was looking into a large polished gem ruby but could not see through it. Mr. Barr's impression of the color at this time was that it was a little more dense than I had described it. . . . By 1905 MST it was apparent that the color was fading.[16]

Greenacre and Barr did not advance any theories on what may have caused the colors they observed, but in a contemporaneous report John Hall, director of the Lowell Observatory, vouched for the authenticity of the sighting. He called Greenacre "a very cautious observer" and noted that Greenacre's boss, William Cannell, "stated that he could not recall that Greenacre had ever plotted a lunar feature which was not later confirmed by another observer."[17]

Thus was reported the first sighting of a lunar transient event, confirmed by two observers and, most important, made by highly qualified personnel. A second sighting by Barr and Greenacre, at the same location, was recorded one lunar month later on November 27, 1963.[18] This observation also was confirmed by Hall and by Fred Dungan, a scientific illustrator on the staff and a qualified telescopic observer. This color feature was reported to be somewhat larger than the one observed in October. It seemed beyond a doubt that something was going on near Aristarchus, since other observers before and after Greenacre and Barr recorded similar activity in the vicinity.

Aristarchus is the brightest feature on the Moon's nearside. This fact, along with the odd shapes of nearby features, suggested that it was of "recent" volcanic origin. (Recent is a subjective term, since no one could then be sure of the relative ages of any lunar features, and the absolute times when they were formed were even larger unknowns.) By USGS's reckoning brightness equated to "young," and these color changes could mean that volcanic processes were still taking place on the Moon. This was an exciting prospect for those of us deciding what experiments to perform on the Moon. Thus, every night that I spent at the telescope I devoted some time to looking at Aristarchus, hoping I would see one of these "eruptions." I never did.

After setting up the contract at the University of Virginia, I contacted an astronomer friend at the NASA Goddard Space Flight Center, Winifred C. "Wini" Cameron, suggesting we start a nationwide network of amateur and professional astronomers to maintain a continuous Moon watch for transient phenomena. Wini was already studying the origin of lunar features and was working with John O'Keefe at GSFC, so this activity fit neatly with her ongoing work. The idea was to publicize a telephone number where people could call in their observations. The person manning the hot line would then contact other observers to try to confirm the sighting. In spite of the acknowledged professionalism of some who had made sightings, many in the small lunar community were skeptical about such events, so we needed to get independent confirmation. We activated the network under Wini's direction in 1965. She went on to study, extensively, lunar transient phenomena and began a program called Moon Blink that developed instrumentation specifically designed to measure and record such transient events.

Lunar transient events had been reported long before the start of the Apollo program, but as might be expected, Apollo aroused great interest in the Moon in amateur and professional astronomers alike. Many more reports of various types of sightings such as color changes, obscurations, and sudden bright spots were made after Apollo Moon landings became the centerpiece of NASA's space program.[19] Up until this time, however, except for Greenacre's sighting, confirmation had never been possible; subsequently there was independent confirmation of several events.

In 1967, after careful analysis of *Lunar Orbiter V* high resolution photographs of the region of Aristarchus, scientists at the Lunar and Planetary Laboratory at the University of Arizona discovered some interesting features at the location of Greenacre's color sightings. They reported that in Schroeters Valley, near the crater named the Cobra Head, they observed a volcanic-looking cone with flow features on its flanks, and that the crater Aristarchus showed evidence of volcanic activity.[20] These discoveries suggested that Greenacre was observing the effects of ongoing lunar eruptions.

The information gained later during Project Apollo and from follow-on studies makes it seem likely that some type of gaseous emission or other surface changes did take place during this time. Some of the color changes reported may have been imagined or caused by terrestrial atmospheric distortion that fooled the observers, but some were almost certainly real events. Astronauts' observations pertaining to lunar transient phenomena are discussed further in chapter 13. For more on the subject, see selected works by Cameron.[21]

3

What Do We Do after Apollo?

Even before we made detailed plans for including science on the Apollo missions, we undertook planning and analysis for missions that would come later. When I joined NASA in 1963, this planning was being done in Tom Evans's office under the name Apollo Logistics Support System (ALSS), implying a program that would come after the Apollo missions but would capitalize on the Apollo hardware then being designed. Post-Apollo programs were given other names in later years as management attempted to get a commitment to continue lunar missions after the initial Apollo landings.

By late 1963, except for the effort that went into the Sonett Report, little had been done to fill the void in Apollo science planning. And many in NASA claimed that no void existed. The Apollo program had only one objective: to land men on the Moon and return them safely. The astronauts would probably take a few pictures, though no camera had yet been selected. They might pick up a few rocks, but tools for doing this were not under development, nor were we designing the special boxes essential for storing such samples on the return trip. A few forward-looking scientists were beginning to think about these concerns, but no one was receiving NASA funds to develop the equipment needed. Post-Apollo planning was an entirely different matter. Tom Evans's office was already spending NASA funds to address what we should do on the Moon after the initial landings. His group and others in Advanced Manned Missions who were looking ahead had initiated studies at the Marshall Space Flight Center (MSFC) that led to the ten-volume MSFC report *Lunar Logistic System*. This effort was directed at MSFC by Joseph de Fries of the Aero Astrodynamics Laboratory, but it included contributions from other MSFC organizations.

In the fall of 1963, less than six years before the first Apollo Moon landing would take place, no timelines had yet been developed to tell us how long the astronauts would, or could, stay on the lunar surface. Payload numbers for the science equipment were not firmed up and varied from the 100 to 200 pounds estimated for the Sonett Report to the "back of the envelope" 250 pounds allotted later. We all assumed it would be difficult to get a larger allocation until all the Apollo systems had been tested and flown and had their performance evaluated. In spite of the many uncertainties and the lack of firm numbers, we took it as given that the landings (number undefined) would be successful and that the myriad Apollo systems would function as advertised.

Our job was not to question any of the Apollo assumptions. Another office in Advanced Manned Missions, under the rubric of supporting research and technology, was responsible for developing alternative ways to ensure mission success. Not only did we assume success, we were charged with expanding the capabilities of the basic Apollo hardware far beyond the original intent. For example, how could we upgrade the lunar excursion module (LEM) to carry a much larger payload than currently planned? How could we extend the time that the command and service module (CSM) could stay in lunar orbit? How could we increase the potential landing area accessible to the LEM (restricted for the first landings to the Moon's nearside, central longitude, equatorial region) so that we could explore what appeared to be critical geological sites far from the planned Apollo landing zone? And would it be possible to land a modified, automated LEM, turning it into a cargo carrier (LEM truck) in order to bring large scientific and logistics payloads to the Moon? All these questions and many more were already under study when I joined the office. (Later in the program the term lunar excursion module was shortened to lunar module, LM, but at this time LEM was still the preferred name.)

The missing ingredient in all this planning was an explanation for why we wanted to stay longer on the lunar surface and why we needed to modify the Apollo hardware to carry bigger payloads. How long should we stay? How big a payload? It became my job to get answers from the ongoing studies. At the end of July 1963, as one of his last actions at headquarters, Gene Shoemaker had sent a letter to Wernher von Braun, the Marshall Space Flight Center director, asking MSFC to suggest what types of scientific activities should be undertaken on the ALSS missions. Verne Fryklund, as Shoemaker's successor at NASA, continued this effort, and I in turn inherited this inquiry when I informally joined his staff.

After meeting Paul Lowman in Fryklund's office, I quickly learned that he shared my enthusiasm about studying and exploring the Moon. Not having been exposed to normal Washington turf battles and jealousies, it seemed quite natural that I ask Paul to work with me informally on some of the projects I had begun. Paul had already made a name for himself by convincing the Mercury astronauts to use Hasselblad cameras on their flights to photograph the Earth's surface. This was no mean accomplishment, since these former test pilots were much more interested in flying and monitoring spacecraft systems than in being photographers. Most of the astronauts eventually enjoyed taking photos, especially when they were published extensively in newspapers and magazines. At that time *Life* had an exclusive agreement with the astronauts to publish first-person accounts of the missions, and a few beautiful full-color photos of the Earth appeared in the articles that followed each Mercury flight. As a result of this success, Paul continued to coach the upcoming Gemini astronauts in photography.

One of the attractive aspects of working at NASA in those early days was that staff members were given great freedom to attack whatever problem they uncovered, without bureaucratic red tape and worry about turf. Paul had originally accepted his temporary headquarters assignment in order to work with Gene Shoemaker, so with Gene's departure, the reorganization of Fryklund's office, and the arrival of Will Foster, the timing was right. Thus we began a long professional friendship that endures today.

By the time I joined Evans's small team in 1963, we already had the results of some preliminary studies on expanding the versatility of the Apollo hardware. The MSFC *Lunar Logistic System* study had examined the hardware then under development for Apollo and documented its inherent flexibility. With what we claimed would be minor modifications, it would be possible to land the LEM at selected sites with no crew on board. Such a LEM could then be a cargo ship carrying as much as seven thousand pounds to the lunar surface, replacing ascent fuel and other equipment not needed for a one-way, unmanned trip. A LEM with this capacity could carry living quarters, large science payloads, or other types of equipment depending on the mission. It seemed that a crew of two astronauts, arriving in another modified LEM and landing close to one or more unmanned logistics LEMs, could spend as much as two weeks on the Moon by either transferring to the earlier-landed LEM or using other payloads that had preceded them.

Similar studies of the CSM showed that it could be kept in lunar orbit long enough to support a two-week lunar stay. In addition, remote-sensing payloads could be carried in one of the CSM's bays to map the lunar surface in various parts of the electromagnetic spectrum, an undertaking that was receiving more and more backing and attention.

Most of my office colleagues were engineers with degrees in electrical, aeronautical, or mechanical engineering and little training in earth sciences. This background was mirrored by NASA's senior management. We decided the best way to convince our bosses that there would be exciting and important investigations for the astronauts to undertake on the Moon (requiring many days and a wide variety of equipment) would be to illustrate these tasks with terrestrial analogies and describe the type of fieldwork and experiments required on Earth to unravel its own history.

Drawing on the Sonett Report and our own knowledge and experience, Paul and I first visited the rock collection at the Smithsonian Museum of Natural History. We borrowed rock samples of various types that illustrated the Earth's geological diversity and the complex geological and geophysical situations we believed would be encountered on the Moon. With visible evidence of how a planetary body (the Earth) had evolved, we developed a rudimentary "show and tell"—a short course in terrestrial geology and geophysics for NASA decision makers—and extrapolated this lesson to the Moon. We hoped our rock collection, along with maps, photos, cross sections, and such, would stimulate their interest and demonstrate that what we were proposing was real and important. We selected igneous, metamorphic, and sedimentary rock samples, later augmented by a few specimens collected at Meteor Crater, Arizona, that showed how a meteorite impact could make rocks look much different than before they were struck. In 1963 so little was known of the physical characteristics of the lunar surface that we felt free to use almost any type of rock to tell our story. Armed with our teaching materials, we put together a half-hour lecture designed around passing out our rock collection to the audience to make particular points and—we hoped—elicit questions. We started with my office colleagues, honed the presentation, and later lectured to senior staff. Tom Evans and E. Z. Gray were impressed with the story we put together. We were ready to take our show on the road and present it along with recent study results confirming that the astronauts might be able to stay on the Moon for two weeks deploying sophisticated science payloads.

On December 23, 1963, after just four months of getting our story together, Evans was asked to brief a prestigious audience: Nicholas E. Golovin, a member of the President's Science Advisory Committee (PSAC), and staff from the Office of Science and Technology (OST). Golovin had been a senior manager at NASA before going to PSAC. He had earned a reputation as a stern, no-nonsense leader in NASA's early days when he chaired a committee to review the Apollo launch vehicle options and became involved in the internal debate on selecting lunar orbit rendezvous (LOR) as the preferred mission mode. Tom was apprehensive about the briefing, which was designed to inform PSAC about our thinking on post-Apollo missions. Ed Andrews and I went with Tom, but because of Golovin's reputation we were told just to listen unless Tom asked us to answer a question.

I thought the briefing went well, and I only responded to a few "geological" questions directed my way. Golovin asked several questions, some in a peremptory tone that I assumed was his normal manner. Donald Steininger, from OST, asked a few questions on classifying rocks, obviously trying to understand how much sampling would be necessary to understand the Moon's history. Tom saw the meeting more negatively. He didn't think we had convinced our audience of the need for extended lunar exploration. As it turned out, Tom's instincts were right: after President Kennedy's death, the Johnson administration never fully embraced post-Apollo lunar exploration.

Of course, not knowing in 1963 and 1964 what events would take place that might dash our plans, we charged ahead and prepared for the big show, a briefing on our vision of post-Apollo lunar exploration for George Mueller, Tom and E. Z. Gray's boss. Mueller, a former professor of electrical engineering, was a slender man with dark hair combed straight back, whose thick, black-rimmed glasses gave him an owlish look. In the meetings I had attended he was soft-spoken and deliberative. I was looking forward to this chance to brief him. Mueller's management style was somewhat unusual compared with that of other managers I had known, and in the years ahead it set the tone for the Apollo program.

After we moved to 600 Independence Avenue (across the street from a parking lot that later was the site of the Smithsonian Air and Space Museum), briefings and status reviews for Mueller were held in Office of Manned Space Flight (OMSF) conference room 425. The room was set up to hold forty to fifty, with Mueller and senior OMSF management seated in the front row before three back-projected screens. A lectern for the presenter was usually placed to the audience's left of the screens. Several overhead microphones let the presenter prompt the projectionist for the next vugraph or slide. Al Zito, a civil servant transferred from the navy, ruled the seas behind the screens. You soon learned that if you wanted a smooth presentation, Al had to understand your needs. With an assistant, he would work the three screens like an orchestra conductor, never missing a beat even if the presenter lost his place or questions disrupted the flow. Al became an OMSF institution. He could have written a funny book about NASA in the years leading up to the first Apollo flights, for he was privy to more senior-level decision making than almost anyone else. Such a book could have included the faults, foibles, and stumbles of many senior managers unprepared for the grilling they got on the stage in room 425.

We had a small art department to develop presentation material for OMSF offices. Housed in the basement of 600 Independence Avenue, it was run by Peter Robinson, who had a full-time staff of six or seven artists and technicians. Pete was a true NASA treasure—unflappable in the face of impossible deadlines yet smiling and friendly and somehow always delivering the goods. I came to know Pete and his team well over the years. I often spent hours in Pete's office along with Jay Holmes, who worked on Mueller's staff to develop presentations, sketching and revising new material for briefing senior management. Mueller had a special ability to make a flawless presentation with minimum preparation before audiences of all descriptions, keeping them spellbound with the colorful and exciting pictures we and others provided. Every program manager soon learned to keep a file drawer full of up-to-date vugraphs of his project, ready at a moment's notice to either give a presentation or provide material for someone else to present.

Although the conference room had microphones to cue the projectionist, there was no way to amplify what was being said for those in back. During and after presentations, Mueller and his staff would ask questions and discuss the matter at hand, with Mueller taking the lead. His voice was soft and low, and since he seldom raised it, even during contentious debates, everyone would be absolutely silent so as not to miss what was being said in the front of the room. In spite of straining to hear, those of us in the cheap seats often could not get the gist of the discussion.

After the meeting we would discreetly mill around in the corridor outside asking "What did he say?" about a particular subject of interest. We usually had to ask two or three people before we got the whole answer, since even those seated closer might not have heard everything. I have often wondered if Mueller knew about these sessions and purposely pitched his voice low to keep everyone focused and eliminate unwanted questions on his time. Whether or not it was a ploy, his meetings usually zipped along, unlike those run by many other managers I have worked with.

The staff had two strategies for briefing Mueller. During the regular workweek we tried to schedule our briefings early in the morning, because as the day wore on, even if you were on his schedule, he would often be called away for urgent telephone calls or for short or long discussions back in his office. His calendar was always filled, so if you didn't finish your briefing in the time allotted it was difficult to get back on his agenda. We quickly learned to schedule important decision-making meetings on Saturday or Sunday, when interruptions were at a minimum and we could talk in a more relaxed environment. NASA Manned Space Flight under Mueller became a seven-day-a-week job, and the lights burned late in most offices at headquarters as we tried to keep up with the rapidly evolving program. The same was true, I know, at the NASA centers.

Our briefing for Mueller was carried out in an atmosphere less formal than usual and with fewer attendees. We made our case for longer staytimes and larger payloads, and since I was at the front for my presentation, this time I had no trouble hearing his questions. Our briefing and props succeeded beyond our expectations; eventually E. Z. Gray felt comfortable enough with our story that he borrowed our presentation for his own briefings, and Mueller soon began to lobby for post-Apollo missions. Over the next two years, as more and more information on the Moon's characteristics became available through new studies and the unmanned missions, we improved our story and eventually made our presentation, without the rocks, at national scientific meetings and symposia.

In the spring of 1964, as we continued to spread the gospel of lunar exploration, Tom Evans scheduled a trip to Houston to discuss our ideas and plans for post-Apollo exploration with some of the staff at the newly formed Manned Spacecraft Center (MSC; later named the Lyndon B. Johnson Space Center). Many of the new arrivals at MSC had been transferred from the NASA Langley Research Center, and one of the more senior was Maxime "Max" A. Faget. Max was a feisty aeronautical engineer who had been a member of the NASA Space Task Group, the source of many of the initial Project Mercury program managers and other senior managers for the fledgling NASA. In 1959 he served on the Goett Committee that recommended increasingly difficult missions, from Project Mercury to Mars-Venus landings, including manned lunar landings. With this background we thought he would be interested in and supportive of our plans. Max's title was director of engineering and development, and as one of the designers of the Mercury capsule he now led the MSC engineering teams responsible for the design of everything from the LEM to space suits.

Tom took three of us with him to Houston to be available for questions from Max and whoever else he might invite to the briefing. At this time the MSC staff was still small. Some members, including Max, were housed in a building near downtown Houston while their permanent offices were being built in a cow pasture at Clear Lake, about twenty miles southeast of Houston. Max brought about six staff members to our briefing, which Tom Evans gave in its entirety. He described in detail the type of tasks we thought would be needed after the initial Apollo landings to answer fundamental questions about the Moon's origin and explained the value of using the Moon as a lunar science base. To carry them out, Tom explained, would require making changes to the projected Apollo hardware so that astronauts could remain on the Moon for weeks at a time and so that large logistical payloads could be carried. As the briefing progressed, there were no questions from Max or any of his staff. Finally, after about an hour of talking, Tom completed the briefing and asked for comments or questions. After a short pause, Max, a short, stocky man with a receding hairline and a bulldog demeanor, turned in his swivel chair and asked in a raspy voice, of no one in particular, "Who thought up these ideas, some high-school student?"

Despite his look of great consternation, Tom calmly tried to explain how we had arrived at our position, but it was clear that Max wasn't interested. Perhaps he had more pressing matters on his mind, such as the first Gemini program launch, which would soon be announced. Perhaps he knew that these ideas were based in part on work done at MSFC, a rival for management of pieces of the Apollo program. The briefing ended in some disarray because of Max's attitude. We quickly left and flew back to Washington, dismayed at our inability to get a more positive response. This was my first encounter with Max Faget and some of the MSC science staff, and it signaled the beginning of a long and often contentious relationship with some MSC offices that lasted until the final Apollo flight splashed down.

No story about NASA would be complete without some discussion of budgets. There have been several accounts, perhaps apocryphal, of how NASA administrator James Webb and his staff arrived at a dollar figure for how much the Apollo program would cost American taxpayers. The most common story had it that his managers told him it would take $12 billion or $13 billion to achieve a manned lunar landing and return, so he made an appointment to discuss the program and budget that he was recommending with President Kennedy. On the way to the White House in his Checkers limousine, a modified version of the popular taxicab (he was the only agency head to use such inelegant transportation, which he found spacious and easy to get in and out of), based on his experience as director of the Bureau of the Budget and his expertise in dealing with big government programs, he doubled the estimate to $25 billion. Whether or not the genesis of this number is true, his projection was on the mark, and the Apollo program eventually was completed for almost precisely that amount.

Webb and his deputy, Hugh Dryden, were the only political appointees at NASA. Webb had been appointed by President Kennedy at the beginning of his term to succeed NASA's first administrator, T. Keith Glennan. Webb was a lawyer who came to NASA from the private sector, but he had been a senior government official in previous administrations and still maintained close ties to important political figures. During his tenure at NASA he was admired for his political astuteness and his ability to move Congress and administrations in the directions he chose. As the Mr. Outside of NASA, he smoothed the way for the agency to grow and prosper during the hectic first years of the Apollo era.

I don't recall any meetings with Webb or Dryden—I was much too junior for such exalted company—but I did attend many meetings over the years with Bob Seamans, the associate administrator and number three man in the management pecking order. His background was very different from Webb's. He had spent most of his career at MIT, first as a professor and later working on a variety of military projects at what was then called the Instrumentation Laboratory. In his autobiography, *Aiming at Targets*,[1] Seamans recounts being recruited by Glennan in 1960 to be NASA's "general manager," running the day-to-day operations. After Webb succeeded Glennan, Seamans continued to fill the general manager's position and became NASA's Mr. Inside. It was in that role that I first met him soon after I joined NASA. I'm sure he wouldn't remember that meeting, and I don't recall the subject (although it probably had something to do with lunar exploration), but I remember one exchange vividly. During the presentations, I asked a few questions. Seamans turned abruptly in my direction and said in a pained voice, "This is my meeting." I may not remember what was covered at the meeting, but those words are etched in my memory. His outburst quickly put a lowly GS-13 in his place, and from that point on I only listened.

Under Seamans's direction NASA quickly became a polished management team. He instituted comprehensive monthly status reviews (general management status reviews) where he presided. Every aspect of all the programs was reviewed, problems were thrashed out, and actions were assigned. It was almost impossible to hide a problem in such a forum, and the business of the agency moved ahead briskly. Eventually Seamans was appointed deputy administrator, and he stayed at NASA until January 1968, the eve of Apollo's biggest successes, for which he could take major credit. In 1974 President Gerald Ford appointed Seamans to lead a new government entity, the Energy Research and Development Agency, and I had the pleasure of working for him again, only this time in a much more senior role.

Only a small fraction of the $25 billion Webb asked for found its way into the Advanced Manned Missions budget or its predecessor offices. It has been difficult, thirty-five years after the fact, to reconstruct these budgets from existing NASA documentation and from my own files. But it appears that from fiscal year 1961 to FY 1968 our offices received about $100 million out of the overall Manned Space Flight budget. These dollars funded a variety of studies: manned lunar and planetary missions, vehicle studies, Earth orbital missions, systems engineering, and other special studies, all related to programs that might follow a successful Apollo landing. In turn, Evans was allocated his small portion of these overall budgets for his office's studies. By FY 1964 he had received a little over $7 million, which he had divided among five competing study areas, and increased funding came our way over the next few years. In the first two and a half years that I worked for Tom and his successors (calendar year 1963 to CY 1965), we had access to about $8 million to start obtaining some hard numbers that would back up the "how long, how big" assumptions for the ALSS missions that we grandly threw around in our briefings and rock lectures. In addition to contractor studies, this funding included a few hundred thousand dollars that was transferred to the United States Geographical Survey (USGS) in FY 1964 and FY 1965, to begin geological and geophysical field studies of how to carry out specific operations during lunar missions with long staytimes. In the early 1960s, you could get a lot of bang for your NASA buck.

My first contractor study was undertaken toward the end of 1963 by Martin Marietta. The company had been in competition with Grumman to build the lunar excursion module, and in the final selection Grumman won. During the competition, Martin had built a full-scale mock-up of its concept of what a LEM would look like. Not surprisingly, since they were both bidding to the same specifications, the Martin concept looked very similar to the winning Grumman model. This mock-up now sat in a high-bay building at the Martin plant in Middle River, Maryland, near Baltimore. Disappointed by the loss, and learning of our activities, a Martin manager came to my office one day to see if there was any interest in using this equipment. Having just completed a parametric analysis of contingency experiments for Apollo, I saw the opportunity to determine, in a preliminary fashion, what difficulties the astronauts might have in making observations from the LEM once they landed on the lunar surface and before they set foot outside. In the back of our minds was the fear that after a successful touchdown something might keep them from getting out on the lunar surface.

Because Martin had the only look-alike version of a LEM, I was able to justify a sole-source contract, and one was soon in place. As part of the contract, Martin did its best, within our funding limitations, to simulate a lunar surface surrounding the LEM mock-up on the floor of the high-bay building. Tons of ashes, sand, and other material were poured on the floor, and we also scattered various types of rocks in the loose, finer-grained material, including some of those we had borrowed from the Smithsonian. To simulate lighting conditions the astronauts might encounter on the Moon, we illuminated the simulated surface with light ranging from low to intense and varied the angle to duplicate the changing sun angles they might confront depending on when during a lunar day they landed.

Since this was to be a simulation of human factors as much as geological conditions, the contract was managed by the Martin human factors department under the direction of Milton Grodsky. The "astronauts" were Martin employees selected by the company. Paul Lowman and I gave them some rudimentary geological training, concentrating on how to make visual observations, provide verbal descriptions using geological terms, and take photographs from the LEM windows to show the nature of the simulated lunar surface. The

Martin test subjects volunteered to spend three or four days isolated in the LEM mock-up, eating and sleeping in the confined space and able to communicate with the test engineers only by radio. The living conditions inside the Martin mock-up, though somewhat uncomfortable, were considerably better than those faced by Neil A. Armstrong and Edwin E. "Buzz" Aldrin Jr. five years later during the first lunar landing and by astronauts in later missions. Armstrong and Aldrin, for example, didn't get much rest during their twenty-hour stay. When they tried to sleep after returning to the LEM from extravehicular activity (EVA) on the surface, Armstrong had to rest on top of the motor casing of the ascent stage rocket, while Aldrin curled up in a confined space on the LEM's floor. Neither slept soundly, and Armstrong perhaps not at all. We were easier on our test subjects; we gutted the interior of the mock-up, and each test "astronaut" had enough space to sleep on a thin mattress on the floor.

The first problem was how to photograph and describe the scene outside the LEM, which had only two small windows, both facing in about the same direction. With this limited view, less than half the lunar surface would be visible if the astronauts could not get out. The LEM also had an overhead hatch to allow them to enter it from the CSM while in lunar orbit, and in that hatch was a small window designed to permit star field sightings, if needed, to update the LEM's guidance and navigation system. But on the lunar surface this window would face only the dark sky above the Moon. The LEM would be equipped with a small telescope that could be operated from inside to assist in the star sightings. We simulated opening the hatch on the lunar surface, with one of the test subjects standing in the opening to make observations. That worked quite well, and we were confident that if this was allowed we could get a good description of the landing site supplemented by panoramic photographs. But what if the astronauts couldn't open the hatch or weren't permitted to do so?

Perhaps we could adapt the telescope—design it to operate more like a periscope so they could scan the surface in all directions. Paul and I traveled to Boston to ask these questions at MIT's Instrumentation Laboratory. The lab had the NASA contract to design the guidance and navigation control system for the CSM and LEM. The telescope was an integral part of the system, along with a sextant in the CSM. We spent the afternoon describing our Martin study and explaining the added value of designing the telescope so it could not only take star sightings but scan the surface and accept a handheld camera to let the astronauts photograph the full surface area of the landing site from within the LEM. The engineers thought this would be possible, but it would entail a major design change to the telescope. Since they were already having some trouble meeting contract objectives, we knew that asking for such a change, based on a perhaps unlikely contingency, went beyond our pay grade. I wrote a short report of our visit and then drafted a memo to George Mueller, for Homer Newell's signature, requesting that modifications to the LEM periscope be considered to permit terrain photography and visual observations of the lunar surface.[2] I have no record of how this request was processed in OMSF, but the modifications were considered too extensive and costly, and the matter was dropped. We resurrected this idea some time later, but again it was not implemented, and fortunately such an instrument was never needed on any of the Apollo landing missions.

With the Martin Marietta contract under way, I started to lay plans for several other studies. The Sonett Report made it clear that we would need a geophysical station of undetermined design that could support five or six experiments. A drill that could extract core samples from deep below the lunar surface was another piece of equipment we believed the scientific community would eventually call for. After studying the first USGS geologic maps of the Kepler and Copernicus regions, traverses of tens of miles seemed necessary if we were to fully understand such large craters, some twenty and fifty miles in diameter. To work far beyond their immediate landing site, the astronauts would have to be mobile, and the more capable we could make a vehicle the more useful it would be. According to our limited understanding of the ongoing designs for the astronauts' space suits and life-support backpacks, they would never be permitted to make such long traverses on foot; they would need a vehicle with a pressurized cab and full life support.

Our growing knowledge of the Moon suggested that the lunar surface might be stable, not subject to shaking and movement. If that was true, it would be easy to design astronomical devices to take advantage of this characteristic, perhaps by using small, symmetrical craters to support radio antennas or large mirrors. With no intervening atmosphere, telescopes operating on the lunar surface during the fourteen-day lunar nights might provide the best "seeing," or "listening," that astronomers could hope to find nearby in our solar system. We proposed to study such instruments for inclusion in the science payloads of these longer missions following the Apollo landings.

Compared with Apollo, where we were told there would be constraints on all the important exploration parameters such as payload weight, surface staytime, and site accessibility, we could think big. The biggest constraint to be removed was the limit on the payload we could send to the Moon's surface. Instead of numbers like 250 pounds, we could plan around payloads of 7,000 pounds or more, which in turn could be used for any need we had. Experiments, life support, and transportation headed the list of items we would try to define so as to take advantage of the larger payloads.

As it was with Apollo, the astronauts' safety was always uppermost in our thoughts as we laid these plans. Other self-imposed criteria required automating as many jobs as possible to conserve the astronauts' time. Lunar surface tasks would be designed to optimize their inherent ability to accomplish those aspects of exploration that humans do best: observing, describing, manipulating complex equipment, and responding to the unexpected. We did not want them performing a lot of manual labor if it could be avoided. But we had to strike a delicate balance between automated functions and manual tasks, or supporters of unmanned exploration, both inside and outside NASA, would raise many questions and objections. Why go to the expense, not to mention risk, of sending astronauts if all they did was turn a switch and let a machine do the work? Switches could be turned on and off from Earth. Our office never thought this was a real challenge, since the astronauts' unique abilities would always be their most important contribution toward exploring the Moon. A combination of automated equipment and hands-on tasks would be needed, and we took it for granted that exploration would proceed in this fashion.

Designing a drill for studying subsurface conditions (called logging) on the Moon and for taking subsurface core samples was a good example of how we eventually applied these criteria. On Earth these operations are labor intensive, requiring many types of laborers and technicians to carry out the wide variety of jobs each entails. Being familiar with all these tasks after spending many months at well sites in Colombia, I could see that new thinking would be required. Terrestrial drilling, logging, and coring equipment must be bulky and heavy to accommodate difficult drilling conditions and the constant rough handling encountered in the field.

Drilling on Earth has one other important characteristic that would be different on the Moon. Water or water-mud mixtures are normally pumped into a drill hole to cool the bit, bring the rock cuttings to the surface, and keep the hole from caving in. Where a water mixture cannot be used, air is circulated under high pressure to accomplish the same purposes. Either of these methods would be impractical on the Moon; we would have to find other ways. Since the primary purpose of drilling on the Moon would be to extract a core, we didn't want astronauts to have to constantly oversee the drilling and coring. This added another dimension to whatever designs would be proposed: a highly reliable, semiautomated lunar core drill. We envisioned much more elegant equipment than that employed on Earth—probably to be used only once at each landing site and thus far different from traditional terrestrial designs.

With all these considerations to be dealt with, the next priority after we started the Martin study was to find a contractor who would do an overall analysis of science needs for the ALSS missions. This new study would generate first-order estimates of weights, volumes, and data transmission and power requirements for a suite of instruments selected by the government. This was my first attempt at writing a government request for quotation (RFQ), and I got help from my office and the NASA headquarters Procurement Office. The RFQ, called "Scientific Mission Support Study for ALSS," focused on the scientific operations that could be done from a mobile laboratory carrying two astronauts. It was released in early 1964 from our headquarters office.

While I was writing this RFQ it became clear that managing contracts from headquarters would be difficult since we had so many studies to get under way. We needed to find a NASA center that would agree to manage them. Also, we reasoned that having a center take ownership of the studies had another advantage. The center would be a strong voice supporting our ideas at other NASA offices that might be skeptical of their importance when budget time rolled around and we were competing for scarce funds.

My few brief encounters with the MSC staff had not been encouraging. They were focused on Gemini and just beginning to think about Apollo science. As shown by our briefing to Faget, planning what should be done after Apollo was not on their agenda. In addition, in early 1964 I could not identify anyone I thought had the right background to manage the studies. Goddard Space Flight Center had built a strong earth sciences staff that could have taken on these studies, but they reported to the Office of Space Science and Applications, the wrong part of NASA. The Kennedy Space Center, although an OMSF center, did not seem to be an option, since its primary responsibility was to service a variety of launch vehicles and there were few earth scientists on the staff. That left the Marshall Space Flight Center, the remaining OMSF center, as my only choice. It turned out to be a most fortuitous final candidate. The studies initiated by our office and others in Advanced Manned Missions to improve the Apollo hardware had been undertaken by several MSFC organizations. Many MSFC staffers had worked on studies reported in the multivolume *Lunar Logistic System.*

Wernher von Braun, a German expatriate rocket genius, was the newly appointed MSFC director. He had just been reassigned from his position as director of the Development Operations Division of the Army Ballistic Missile Agency at the army's Redstone Arsenal, located with MSFC in Huntsville, Alabama. At the end of World War II the army had brought more than 120 German engineers and scientists, led by von Braun, to the United States to improve the country's rocket know-how. Some of this original group had been assigned to Cape Canaveral as well as Huntsville. With a perfect launch record for their rocket designs, they successfully launched the first United States satellite, and our rocket technology was progressing rapidly. Sending men to the Moon was to be their next challenge, which would include building the huge new *Saturn V.* MSFC was NASA's largest center in terms of manpower, so the question became where to go in this organization, with which I had had no previous contact. The decision turned out to be easy, since the Research Projects Laboratory (RPL), under Ernst Stuhlinger, one of von Braun's original team members, had been responsible for writing volume 10, *Payloads,* of the *Lunar Logistic System* report.[3] This volume described science payloads that could be carried on modified Apollo spacecraft, including many geophysical experiments.

After several phone calls I scheduled a meeting with James Downey, manager of the Special Projects Office in RPL; he and some of his staff had also contributed to volume 10. Our first meeting took place in late 1963 and was marked by some careful bureaucratic dancing. Reflecting his center's and his immediate boss's cautious, Germanic approach to having someone from headquarters ask for a commitment of manpower and center resources, Jim wanted to know if my request represented a formal headquarters assignment of new duties for MSFC. I wasn't prepared for such a pointed inquiry and knew I didn't have the authority to say yes, so I hedged but assured him that our office had funds to support the studies I was asking him to manage.

Jim, a University of Alabama graduate, was an easygoing manager who commanded the respect of his unusual, multitalented conglomeration of scientists and engineers. He was eager to take on this new job, for so far his office had not received much funding for its studies. An important measure of a successful manager at NASA was how much funding he obtained and how many contracts he managed, so the promise of new funding was well received. But before he could agree it would have to be formally requested through the proper channels. From my brief exposure to his staff, it appeared that they had the mix of skills needed to monitor the wide range of contractor studies we wanted to perform. I told Jim I would go back to Washington and start the paperwork. This meeting was the beginning of a long and productive relationship with Ernst Stuhlinger, Jim Downey, and their staffs as we undertook several studies that broke new ground for lunar exploration.

What did it mean when a NASA center managed programs or studies? There were many responsibilities. We met frequently to plan future procurements to be sure we all agreed on what the final products would be, and we would estimate the funds required and the schedules to be met by the contractors. Then MSFC would write the request for proposal (RFP), designate a contract monitor on Downey's staff, establish a rather informal source selection committee to evaluate the proposals, advertise the procurement in the *Commerce Business Daily*, release the RFP, evaluate the proposals received (with the evaluation documented in case of a protest from a rejected contractor), choose a winner or winners, award the contract, and then—the important part—monitor the contractor's performance until the job was completed. The procedures we followed for these smaller contracts, although spelled out in NASA regulations, were nowhere near as precise as today's requirements, which call for formally appointed source evaluation boards and source selection officials. Without this time-consuming bureaucratic red tape, we were able to move ahead quickly on our contracts.

In my mind the steps named above more than justified asking a center to help get the contracts under way; the centers had much more manpower available for this cradle-to-grave job, as well as experience in directing the efforts of NASA's growing number of contractors. The main responsibility of NASA headquarters staff was to develop the big-picture programs and run interference with the administration and Congress on issues pertaining to budgets and policy, leaving the details of running the programs to the centers. In reality these distinctions weren't so clear-cut, and the centers and headquarters worked together on all aspects of the programs. Contract management of advanced (paper) studies migrated more and more from headquarters to the centers. As NASA matured as an agency, the centers became powerful independent entities, supported by their homegrown political allies in Congress and the executive branch. This growing independence was one of the reasons friction developed between headquarters and MSC. Under von Braun, MSFC accepted headquarters direction more graciously; perhaps this smoother relationship was a reflection of MSFC's confident corporate personality, embodied in the person of its director and enhanced by its established reputation in rocketry. MSC was the new kid on the block, attempting to prove that it knew how to get the job done but with a short track record. And it had no one with a reputation like von Braun's to intervene if problems arose. Little by little, of course, MSC established this track record with the successful completion of the Mercury and Gemini programs, but this newfound confidence never translated to a smooth management relationship with our headquarters office in matters dealing with science.

Once MSFC agreed to manage our post-Apollo science studies, events moved rapidly. Contracts were signed in 1964 for the studies mentioned above, and soon afterward management of the ALSS Scientific Mission Support Study, won by the Bendix Aerospace Systems Division, was transferred to MSFC. Not all headquarters managers followed this practice; some liked to maintain control of their programs by doing the day-to-day management. But the advantages of leaving contract management to MSFC were evident from the start. Small study contracts could be managed by headquarters staff, since they resulted only in paper, but once prototype hardware became deliverable, only a center could supply the management expertise and resources needed. Several of our contracts required delivery of engineering models or "breadboards" of proposed equipment as well as detailed analyses.

In June 1964, along with some reorganization at headquarters, the ALSS program was modified and given a new name, Apollo Extension System (AES). The new name was meant to convey a different message than Apollo Logistics Support System; AES was to be a new program based more closely on Apollo but not requiring the extensive hardware modifications envisioned for ALSS. There would still be a greater potential to study the Moon, both on the surface and from lunar orbit. We could still plan on dual launches of an automated LEM shelter-laboratory and a LEM taxi to carry the astronauts to the surface and return them to rendezvous with a CSM built for extended staytime. Our strategy, as we had planned for ALSS, centered on the astronauts' transferring to a shelter-laboratory after landing and conducting their extravehicular activities from there. AES studies also included using a wide variety of instruments aboard the Apollo CSM in Earth and lunar orbit to survey and map the surfaces of these two bodies. The orbital studies would now be managed in the Advanced Manned Missions office as a continuation of the work initiated earlier by Pete Badgley.

In early 1964, President Johnson asked NASA to develop long-range goals for the agency and, by implication, the nation. Homer Newell, as was the custom, quickly asked the National Academy of Sciences to help provide a response focusing on space science. In 1961 the Academy's Space Science Board (SSB) had recommended that "scientific exploration of the Moon and planets should be clearly stated as the ultimate objective of the U.S. space program for the foreseeable future." Now, three years later, Harry Hess, chairman of the Space Science Board, wrote to Newell indicating that a change in objectives was appropriate. Planetary exploration, starting with unmanned exploration of Mars and eventually leading to manned exploration, should be the new goal.[4] The SSB stated that Mars "offers the best possibility in our solar system for shedding light on extraterrestrial life." It was ready to concede that the Apollo program would be successful, thus the new emphasis on planetary exploration. But the SSB also suggested some alternatives that included extensive manned lunar exploration leading to lunar bases. These recommendations, which we took as an endorsement of the studies we were pursuing, were eventually incorporated into the report that was sent to the president. In the fall of 1964 we believed our programs would soon be officially embraced by the administration, and this belief was reinforced a few months later when the president publicly declared that "we intend to not only land on the moon but to also explore the moon."[5] We waited in vain for a formal start. Instead Johnson focused on his Great Society programs and, increasingly, on the war in Vietnam. There were three more years of growing budgets for Manned Space Flight to fulfill the lunar landing mandate, but NASA's overall funding peaked in FY 1965 and thereafter began to decline.

At the end of 1964 Ed Andrews and I were transferred from Tom Evans's office to a new office called Special Studies under the direction of William Taylor. I was not pleased with this move; the mission of this new office was poorly defined, and it removed me from the day-to-day oversight of the programs I had initiated. I maintained contact using my other hat, however, working for Will Foster. Evans was promoted to lieutenant colonel that summer, and soon he left NASA and the army to return to Iowa and manage his family's large farm. With his departure, the Advanced Manned Missions Lunar and Planetary Offices were combined under Frank Dixon, who until then had been director of the Manned Planetary Missions Office.

In June 1965 I was transferred back to Manned Lunar Missions Studies, once again a separate office, under a new director, Philip Culbertson, brought in from General Dynamics to replace Evans. I mention these office moves only to illustrate the uncertainty that was present at NASA as top management tried to position the agency for life after Apollo. Although Manned Space Flight's budgets were still growing, management could foresee that if new missions were not assigned soon, the agency would be largely marking time until the end of Apollo. The mantra in OMSF was that only large, manned-mission programs could sustain NASA. Other programs, such as unmanned space science and aeronautics research, though important, would never maintain a prominent agency in the federal government's hierarchy, which consists of large cabinet-level departments and also smaller independent agencies like NASA. In Washington, big, growing government programs were good for those managing them, and declining budgets were bad for ambitious managers.

At the same time as we were attempting to define the science content of the ALSS-AES missions, the Boeing Company's lunar base study, with the title Lunar Exploration Systems for Apollo (LESA), was under way. When William Henderson joined our office at the end of 1963 he became the headquarters lunar base expert and assumed oversight of all the lunar base studies. Boeing's final LESA report described a modular lunar base that would be assembled from Apollo hardware, incorporating greater modifications than required for ALSS-AES missions. By grouping modules, a base could support colonies of two to eighteen men. (We had no women astronauts at that time, so the studies were always described in masculine terms.) Individual modules might take as much as 25,000 pounds of useful payload to the lunar surface. Depending on the mix of equipment and the number of modules, these colonies could operate for ninety days to two years. We envisioned sending to the Moon large pieces of scientific equipment that would permit a wide range of activities. Long-duration geological and geophysical traverses in large wheeled vehicles could be conducted, as well as studies confined to the base, such as deep drilling and astronomical observations. These endeavors, we believed, would lay the groundwork to justify permanent bases.

During this period we persuaded our management to let us take several trips overseas to gain greater insight into some of the situations we expected to encounter during lunar exploration. In January 1964 Bill Henderson took the first of such trips, receiving permission to visit our scientific bases in Antarctica. He made the case that these stations were the closest examples we could find to what a base on the Moon would be like: isolated, difficult to supply, and therefore self-sufficient. Their primary reason for existence was to conduct scientific investigations; the secondary objective was to show the flag—or perhaps vice versa. Both these reasons closely followed what we believed would be the ultimate rationale for establishing lunar bases, and one couldn't deny that Antarctic conditions were moonlike. Bill thought his time in Antarctica was well spent and, since he was the only person at headquarters with this experience, his recommendations carried more weight when he advanced his thoughts on how to design a lunar base.

At the end of the rather massive Boeing study, Bill initiated a new round of more detailed lunar base analyses. The resulting contract, signed by the Lockheed Missile and Space Company in February 1966 for $897,000, was the largest award ever made by our office. The study, called Mission Modes and Systems Analysis, would be supported by three other contractor studies valued at an additional $900,000. One of these studies, Scientific Mission Support Study for Extended Lunar Exploration, was won by North American Aviation, with Jack Green, of the "volcanic Moon," playing a prominent role in the study. The contract would be monitored by Paul Lowman and Herman Gierow, Jim Downey's deputy and a versatile manager who had participated in the earlier LESA studies.

For decades space dreamers and enthusiasts, including MSFC's director, von Braun, had written and lectured on the possibility of establishing a lunar base. Now major government funds were to be spent on a serious look at what it would take to carry it off. The inherent ability of the Apollo hardware to place large payloads into Earth orbit and send them on to the Moon was the initial requirement for lunar base planners. After modifications, with each flight the Apollo upper stages would be capable of placing large payloads on the lunar surface. Big payloads meant you could envision supporting and supplying a large lunar colony over long periods at a reasonable cost. This was the challenge, first to Boeing, then to Lockheed and its support contractors: Tell us how it could be done, what such a base would look like, and how a base could support scientific and engineering operations that would justify its existence. The results of all these studies were encouraging, especially assuming that the nation would continue to commit large amounts of money to the investment it was making in Apollo—not an unreasonable expectation in the mid-1960s. Extended lunar exploration, followed by the establishment of one or more lunar bases, would not be cheap. But the initial analyses seemed to show that, for an additional investment approaching what would be spent on Apollo, all this could be done.

Bob Seamans, George Mueller, and E. Z. Gray began to lobby Congress for a NASA mandate that would implement these grand designs. When they testified before NASA congressional oversight committees, they would impress the members with realistic artists' renditions of what these stations and bases could look like. They also had funding estimates (supplied from our contractor studies) to support their contention that continued lunar operations were feasible at a reasonable price and would produce important results. At a lower level in the management chain, staff like me, Paul Lowman, Bill Henderson, and others involved in the studies at MSFC took every opportunity to advertise our plans at professional conferences and public forums. We could usually count on good coverage from the media, and it seemed at the time that we were winning public support. Public polls always gave NASA high marks, and the major news and trade magazines were eager to write stories and show drawings of future lunar colonies.

Contractors who won our awards usually included well-known scientists on their teams as consultants (a few with Nobel credentials); they were to review study results during the contract and make recommendations to the contractors to ensure that the results were grounded in scientific reality. During proposal evaluations, the quality of these consultants could determine which contractor would receive the award. While the contract was under way, or at its conclusion, we were not bashful about dropping their names if our assumptions were challenged.

Returning to the ALSS-AES studies, in May 1964 MSFC put together the RFP for what we called the Emplaced Scientific Station (ESS). This study would provide a preliminary design of a self-sufficient geophysical station to be deployed by the astronauts on the lunar surface, incorporating several experiments listed in the Sonett Report and some from other sources. We received eight responses to the RFP and selected two contractors, Bendix Corporation, led by Lyle Tiffany, and Westinghouse, led by Jack Wild. These two contracts, along with the Scientific Mission Support Study, would provide us with enough detail that one year later we could extrapolate the results to design the Apollo geophysical station, which would have to meet more stringent requirements.

As we did for the ESS, we awarded two contracts in 1965 to study competing designs for a hundred-foot drill. One went to Westinghouse Electric Corporation and a second to Northrup Space Laboratories. Each contract had a value of more than $500,000. The MSFC contract manager was John Bensko, a geologist who had worked in the oil and coal mining industries before joining NASA. After coming to MSFC, he helped develop engineering models of the lunar surface, useful background for his drill contracts. John put together an advisory team from the Corps of Engineers and the Bureau of Mines to provide additional engineering expertise as the contractors began to cope with their difficult assignments. In those days NASA always attempted to at least match the contractors' expertise in house so that our oversight and evaluation of their performance were well grounded. I believe this respect for each other's abilities let NASA and its contractors work together better as a team, although some contractors grumbled at the tight monitoring. Today NASA's approach to contract monitoring seems to have changed almost 180 degrees; in-house expertise in the aspects of a contract is often minimal. For the drill studies, NASA's competence was especially important, since we planned a series of difficult tests including drilling in a vacuum chamber at MSFC, never before attempted with a drill of this size.

Considering the unusual location for a drill rig and other constraints, the Westinghouse approach to drilling on the Moon was relatively straightforward, modeled after terrestrial wire-line drilling. Short sections of drill pipe were added from a rotating dispenser as drilling progressed; the core would be extracted from a short core stem after each section was taken from the drill hole. Since this would be close to a conventional design, it would entail almost constant monitoring by the astronauts. The Northrup design was radically different. It proposed using a flexible drill string, wound on a drum, that would be slowly fed into the hole to the final target depth of one hundred feet. A core stem would be attached to the end of a flexible pipe, and the core would be recovered much as in the Westinghouse design but without adding drill pipe sections every five to ten feet. Several innovative concepts were aimed at reducing the astronauts' involvement, and though we recognized that they posed some design risks, we accepted them as the price for a possible breakthrough in technology.

One of the major challenges for both concepts was cooling the bit during drilling to reduce wear. Bensko hired Arthur D. Little to do a separate analysis of how to accomplish the cooling. The company's study showed that the cooling problem could be greatly mitigated in the vacuum environment of the Moon if the rock cuttings could be rapidly moved away from the bit face so that the they would carry off some of the heat. Spiral flutes were thus incorporated on the outside of the drill string, like an auger, to lift the cuttings up through the hole to the surface.

Although the spiral flutes partially solved how to cool the bit, as our studies progressed we found that after a short time the bit would still get too hot, become dull, and stop cutting. Both contractors settled on using diamond-core bits to ensure that they could drill through any rock type encountered. Westinghouse had included Longyear on its team, and Northrup had teamed with Christianson Diamond Bits, the leading industrial suppliers of diamond-core bits. Both bit contractors concluded that, with the technology then available, even a diamond-core bit would need to be replaced many times in drilling a hundred-foot hole. This was unacceptable.

Initially, the best the Westinghouse team could do under test conditions was to drill fourteen inches through basalt, a possible lunar rock type, before an uncooled bit failed. But they reexamined the problem and finally hit on a solution. The diamond-core bits then offered to industry used a matrix that "glued" tiny diamonds to the bit in a random alignment. The random alignment did not allow each diamond to present its best cutting edge to the rock being cored, however. They demonstrated that carefully setting the diamonds in the matrix significantly prolonged the life of the bit. Hand setting each diamond would add greatly to the bit's cost, but it would be well worth it for a lunar mission where the astronauts' time was more precious than a diamond bit. These newly designed bits lasted more than ten feet before they dulled. After other design changes, eventually we expected to drill the entire one hundred feet with just one bit, eliminating a time-consuming chore. As I recall, Christianson developed a relatively inexpensive technique to manufacture bits of this design for their terrestrial customers. Although they cost more than normal diamond-core bits, they were worth the investment because fewer were needed.

The cost of drilling on Earth is strongly influenced not only by the price of bits but by the time needed to extract a dulled bit from the drill hole, change bits, and resume drilling.

As the studies continued, progress on the Northrup design slowed, and the contract was terminated before they delivered a complete working model. Our gamble had failed. A Westinghouse model was tested at MSFC, including vacuum chamber tests. Finally tests were held in the desert in Arizona and New Mexico to simulate drilling under lunar conditions (but not in a vacuum), with no lubrication for the bit. Bensko recalls that we chose a bad time for our tests: there had been more rainfall than normal, and the wet soil gummed up the flutes. In other tests the fluted drill pipe performed about as expected, and we were encouraged to believe that a full-scale drill could extract cores on the Moon to depths of one hundred feet.

In anticipation of drilling a deep hole on the Moon, in 1965 we started two studies with Texaco and Schlumberger to design logging devices that would determine conditions beneath the lunar surface. (Taking measurements in terrestrial drill holes is standard practice for obtaining information on subsurface conditions.) These contracts, also worth more than $500,000 each, were managed by MSFC's Orlo Hudson.

In both terrestrial drilling and drill-hole logging, the drill hole is almost always filled with a fluid, of varying chemistry, the remnants of the drilling mud. Lacking this liquid to couple the logging tools to the subsurface rock formations, the contractors were forced to modify standard oil field technology. The Texaco team, which had extensive experience in developing logging devices for oil field exploration, had won an award from the Jet Propulsion Laboratory (JPL) to provide logging devices for the Ranger and Surveyor projects. In their planning stages both projects included small drills as potential science payloads. Schlumberger, the acknowledged leader in developing logging devices for the oil and mineral exploration industry, showed an interest in such unworldly studies (to our surprise), entered a bid, and won the other contract. Both contractors overcame the lunar logging constraints and designed a suite of devices that could make measurements in a hole drilled on the Moon. Perhaps one day, when the opportunity arises to drill deep holes on the Moon or some other extraterrestrial body, these studies will be found and reread.

The most interesting set of studies we conducted were those related to providing mobility once the astronauts reached the lunar surface. Many concepts were being proposed, some more fanciful than others. MSFC had reported the results of the first in-house mobility studies in volume 9 of the *Lunar Logistic System* series.[6] Two of the main contributors to these studies were Jean Olivier and David Cramblit, who wrote several reports on lunar surface mobility. To learn what types of mobility systems would work best on the Moon, based on the limited knowledge available, MSFC and the Kennedy Space Center developed a lunar surface model to study how wheeled vehicles might perform on soils in a lunar vacuum and what type of obstacles they would have to traverse.[7]

JPL had also developed a lunar surface model in order to design a small unmanned vehicle for the Surveyor project.[8] It had tested several designs on simulated lunar terrain in the early 1960s. My first trip to JPL was to witness a test of a small vehicle operated by an engineer with a handheld remote-control box, hardwired to the rover. It was much like a modern toy car except for the connecting wire. Today's electronics permit cheap radio-controlled toys; in the early 1960s radio control was a luxury we usually did without when testing our concepts. This was an interesting demonstration of a small articulated vehicle with springy wheels driving over loose sandy material and small rocks. From time to time there were short interruptions caused by failures in the then state-of-the-art electrical circuits, powered by vacuum tubes. One could say that the granddaughter of this vehicle was the small rover named *Sojourner* that traversed the Martian surface in July 1997. A United States automated rover never made it to the Moon, but a Soviet rover named Lunokhod operated on the Moon in 1970.

Although in 1964 and 1965 we still did not have any data from direct contact with the lunar surface, information from radar and laboratory studies predicted how the Moon's surface layer would respond to a wheeled vehicle. In spite of Tommy Gold's theories, we were certain that a vehicle could move around without serious difficulties. But we were not sure how the Moon's almost total vacuum would affect the lunar soil; the high vacuum that would be encountered on the Moon was impossible to achieve on Earth. Studies had been conducted in high vacuum using several types of simulated lunar soil, but their fidelity was open to question because our ideas about the composition of lunar soil (grain size, mineralogy, and other characteristics) were mostly guesses.

Our first contractor studies of a lunar surface vehicle were undertaken by the Bendix Corporation and the Boeing Aerospace Division. They were selected in

May 1964 to study ALSS exploration payloads, including a vehicle we had dubbed MOLAB (for mobile laboratory). The Boeing study was managed by Grady Mitchum, and the Bendix manager was Charles Weatherred. Because of their involvement in the post-Apollo studies, both these men and their companies would be important contributors to later Apollo contracts. Bendix had earlier won one of the JPL design contracts for a small Surveyor rover, so it was well prepared to undertake the study. From taking part in our lunar base studies, Boeing had a good background that included designing mobility concepts.

The concept for using a MOLAB was to have it delivered to the Moon by an ALSS automated LEM. It would then be deployed and operated remotely so that it could travel to another LEM carrying two astronauts that would land a short distance away. It was to be a vehicle of about seven thousand pounds, including the scientific equipment it would carry. It would support two astronauts for up to two weeks in a pressurized cab, permitting shirt-sleeve working conditions while under way. Based on our study of early geologic maps of the Moon, we felt that such a vehicle should have a traverse range of several hundred miles so the astronauts could make several trips far enough from their landing site to sample geologically interesting areas. These requirements were a tall order for any vehicle, not to mention one that must function on the lunar surface.

The two contractors were also asked to design a shelter that could be delivered by the same type of automated LEM and a smaller, unpressurized vehicle we named the local scientific survey module (LSSM). (Moon vehicles had to have strange names; they couldn't just be called cars or trucks, since they would be so different from any of their terrestrial cousins.) All these studies were to be accomplished by both contractors for a total of slightly more than $1.5 million.

As the studies progressed, under the direction of Joe de Fries and Lynn Bradford at MSFC, the MSFC Manufacturing Engineering Lab built a full-scale mock-up to evaluate such things as cabin size and crew station layout. Many photographs of this rather unusual looking vehicle were circulated to the media and other interested groups, showing our progress toward the next step in lunar exploration. A December 1964 issue of *Aviation Week and Space Technology* featured a front cover picture showing the mock-up sitting on top of a LEM truck and included a special report on the Bendix version.[9] The MOLAB, more than any other project we worked on for post-Apollo missions, seemed to catch the imagination of futurists, perhaps reflecting the national love affair with the automobile. Perhaps people could visualize themselves speeding across the lunar surface, dodging boulders and craters.

At the conclusion of the initial contracts in July 1965, both contractors were given extensions totaling more than $1 million to refine their LSSM designs. Bendix and General Motors received two other contracts to produce four-wheel and six-wheel LSSM test designs, each worth almost $400,000. By the end of 1965 we had awarded lunar vehicle contracts for more than $3.5 million and had probably spent almost as much for in-house civil service workers and contractor support.

While all this wheeled-vehicle planning was under way, Textron Bell Aerospace Company was quietly developing a small manned lunar flying vehicle (LFV). A one-man version was demonstrated in a live test early in 1964. (A later generation of this device was demonstrated at large gatherings including the 1984 Olympics in Los Angeles, and a version was flown in the James Bond movie *Thunderball*.) Bell had conducted a preliminary study of how to combine the MOLAB and the LFV, sponsored by NASA's Office of Advanced Research and Technology. In these early days we had a good working relationship with OART; under the direction of James Gangler, it was attempting to look far ahead at technology needs for lunar exploration and lunar bases. After the impressive one-man flight demonstration, MSFC awarded Textron Bell a follow-on contract in August 1964 to further define the concept. In these studies the LFV was given two functions—to return the astronauts to a base camp in case of a MOLAB breakdown and to help them reach difficult sites.

The MSFC contract with Textron Bell called for an LFV design that would carry two astronauts a minimum of fifty miles for the safety fly-back mission. This would also be a useful range to take the astronauts to sites they could not reach overland. MSFC later awarded Bell a second contract with a more modest goal—to support AES missions requiring an operations radius of only fifteen miles. This vehicle, which needed far less fuel because of its shorter range, could carry one astronaut and three hundred pounds of equipment or transport two astronauts the same distance. Both design studies and a working prototype indicated that an LFV with these characteristics was feasible.

A study was also done to assess the advantages of using the lunar surface for astronomical observations, an application supported by some, but not all, in the astronomical fraternity. In 1965 MSFC awarded Kollsman Instrument Corporation a one-year contract for $144,000 to assess the feasibility of carrying a large optical telescope observatory to the Moon mounted on a modified automated LEM lander. MSFC's contract monitor was Ernest Wells, an amateur astronomer whose avocation served him well in this job. Kollsman was already developing the Goddard Experimental Package (GEP), an automated observatory scheduled to be launched in 1966 on the Orbiting Astronomical Observatory (OAO), so working with the company would save effort and money.

The GEP consisted of a thirty-six-inch reflector telescope, its mounting, a camera, and associated electronics. Improvements to the GEP design to take advantage of its lunar location could be recommended during this study, as well as design changes to accommodate the astronauts' involvement in its operation, since the OAO design was a fully automated observatory. The results were encouraging, indicating that the astronomical payload could operate on the Moon for long periods in both an unmanned and a manned mode.[10] Kollsman also reported that new technology, by greatly reducing the overall weight, might permit a much larger instrument, perhaps up to 120 inches in diameter, to be carried on the same LEM truck.

A fallout of these studies at MSFC was the establishment of a Scientific Payloads Division in Stuhlinger's Space Sciences Laboratory. Jim Downey became the director of this new division, and Herman Gierow was named deputy. Later, as the MSFC work on post-Apollo science wound down, both Jim and Herman went on to manage important new programs that included work on the Apollo telescope mount flown on Skylab. Their work on space-based astronomy culminated in the launch of three high energy astronomical observatories in the 1970s and studies of a large space telescope that evolved a few years later into the successful Hubbell space telescope program.

The transition from planning ALSS missions to planning AES missions was relatively painless. AES payloads would be smaller than those we anticipated for ALSS missions but much larger than Apollo's allocation. By this time we had a much better understanding of the Apollo hardware than when we started our ALSS studies, and we were also becoming aware of the potential Apollo operational margins that could permit larger payloads or increase flexibility. We hoped these margins would soon be available as confidence in Apollo's performance grew.

Removing the ascent propulsion and other unnecessary systems required during a normal LEM ascent and rendezvous would free up space for approximately 6,000 pounds of payload, 1,000 pounds less than the total used for the

ALSS studies. Of the 6,000 pounds, 3,500 would be required for consumables and other additions so two men could stay in the LEM for two weeks. The remaining 2,500 pounds could then be used for scientific equipment. This represented a rather firm increase of an order of magnitude over the expected allocation for Apollo science payloads. Although 2,500 pounds was less than half the weight we had been using in planning, it was enough to be exciting.

Based on 2,500 pounds and results coming in from our ALSS-AES studies and USGS work at Flagstaff, we divided a typical payload as follows: 1,000 pounds for a fully charged LSSM with a range of 125 miles, 200 pounds for a hundred-foot core drill, 90 pounds for logging devices, 350–400 pounds for an ESS, 80 pounds for a small preliminary sample analysis lab, 100 pounds for geological field mapping equipment, 150 pounds for geophysical field survey equipment, 30 pounds for sample return containers, and up to 500 pounds for a power supply for the drill or other exploration equipment. We felt this equipment would let the astronauts take full advantage of a two-week stay and study their landing site in some detail. For safety reasons, during manned operations the LSSM would be restricted to a radius of five miles, but it could operate in both manned and automated modes. After the astronauts left it could carry out investigations farther from the landing site, to the limit of its battery charge, under command from Earth.

Our planning for lunar exploration after the initial Apollo landings was now in high gear. The next step was to test our ideas as realistically as possible so we could not be accused of offering proposals thought up by "some high-school student."

The United States Geological Survey
Joins Our Team

At the same time we were conducting our studies at Marshall Space Flight Center, we began to build a strong partnership with the United States Geological Survey under the direction of Eugene M. Shoemaker at Flagstaff, Arizona. Gene, an outstanding scientist, colleague, and friend, had a major impact on the program. I will be discussing his contributions in future chapters. *To a Rocky Moon*, by Don E. Wilhelms, provides many details of Shoemaker's remarkable career; I also recommend this book if you want to read more on Apollo lunar science.[1]

After leaving Washington in the fall of 1963, Shoemaker returned to Flagstaff, where he had recently moved with his wife, Carolyn, and three small children. He had chosen Flagstaff for his new office location for several reasons. It had a small-town atmosphere, and there were many Moon-like geological features only about an hour's drive or less to the east. Another plus, although Gene might have denied it, was that Flagstaff was far enough away that he would be left pretty much on his own, undistracted by his superiors in Washington. But the local geology was the real magnet. Meteor Crater, whose origin Gene had helped unravel, was about to become a star in the geological firmament, a place all the astronauts would visit and study. He may have thought the Branch of Astrogeology would go quietly about its business, but its notoriety was to grow as its close relationship to the astronauts became known.

Although Gene was in Washington for about two months after my arrival, our paths had not crossed. It soon became clear that he was someone I had to meet. As our contract studies progressed and I learned about his work, it seemed there might be a good match between his interests and my office's future needs. His staff was already heavily involved in NASA work, including some projects that could contribute directly to our studies. We talked several times on the phone about the direction post-Apollo planning was heading and agreed to meet and see if we could find areas of shared interest.

My first trip to Flagstaff was in March 1964. In those days the best way to get there from Washington was to catch a late afternoon United Airlines flight to Denver and connect with Frontier Airlines for a milk run to Flagstaff. Frontier had recently started operations as a feeder airline connecting many small western towns with larger cities such as Phoenix, Salt Lake City, and Denver. At this time it mostly used the Convair 240, a two-engine propeller plane. As a passenger carrier, it offered basic transportation, noisy and drafty. The crew consisted of pilot, copilot, and one overworked stewardess attending to the needs of thirty or forty passengers, a few usually sick from the bumpy ride. Since there were frequent stops at cities such as Colorado Springs and Farmington, New Mexico, the plane never reached high altitudes; it flew just high enough to clear any mountain peaks. So you bounced along, buffeted by the thermals that swirled over the mountains below or the clouds above.

On summer trips you dodged thunderheads and lightning all along the flight path and imagined how rough the landscape below would be in a forced landing. By the time you left Denver in the winter it was dark, so all you could see out the small windows were a few lights from the scattered towns below. At some of the small airfields the nearby peaks, unseen in the darkness, towered above the landing approach path. Flagstaff's airport, cut out of a stand of ponderosa pines, was just a few miles south of town and near one of those towering peaks, Mount Humphrey (12,670 feet). As I walked down the stairs at Flagstaff on that first trip, I inhaled the aroma of the ponderosas, unlike any forest smell I had ever experienced. It was a crystal-clear, cold night with no sky glow from the nearby city. At seven thousand feet, the stars were the brightest I could remember since my days at sea. It was easy to understand why Percival Lowell had established his famous observatory near Flagstaff.

Flagstaff had grown up as a two-industry railroad town, serving lumber and cattle. The main street stretched for several miles along old Route 66 (also U.S. 40), paralleling the railroad tracks. Now it was mostly a tourist town, a stop along the road to the Grand Canyon, about eighty miles to the northwest. The Grand Canyon, like Meteor Crater, would become an astronaut training site. Flagstaff boasted a small college, with a few thousand students at that time, and several motels, small restaurants, and tourist shops, most with a western or Native American motif. East of town were Sunset Crater and other volcanic features, and continuing east you could drive through portions of the Hopi and Navajo Indian reservations and the Painted Desert.

The next morning Donald Elston (Gene's deputy—his real title was assistant branch chief) picked me up at my motel and drove me to their temporary offices on the grounds of the Museum of Northern Arizona. Gene met me there, dressed in blue jeans, a western shirt, field boots, and bolo tie—the standard uniform for his staff, although a few were not so nattily turned out. My typical Washington uniform of suit, white shirt, tie, and dress shoes drew some wise-cracks, dictating a change of wardrobe for my next visits. Gene's offices, in several one-story cinder-block buildings, were not imposing. Furniture was rudimentary and looked like army surplus. Some of the more innovative staffers had built bookcases out of packing boxes, and recently Gordon Swann reminded me that when he first arrived in Flagstaff the only extra chair in his small, shared office was a short plank he laid across his wastebasket. In spite of appearances, you could feel the energy and dedication of the staff Gene was putting together; they hadn't come to Flagstaff for fancy accommodations.

Gene introduced me to those present—mostly young, some of them recent college graduates—and gave me a short tour. Gene had been selected as a coinvestigator for Ranger and the upcoming Surveyor program. Some staffers were busy analyzing the first Ranger close-up pictures, returned only four months earlier, and preparing for the first Surveyor landing. In addition to the Ranger and Surveyor work, his office had the lead in making the lunar pho-togeologic maps that would be influential within a few years in the selection of potential Apollo and post-Apollo landing sites. Most of this latter work, sup-ported by Bob Bryson at NASA headquarters, was being done at the branch's offices in Menlo Park, California, using the nearby Lick Observatory telescope. Several Flagstaffers commuted to California to work on their assigned quad-rangles; Gene had tried to get as many of his staff as possible involved in the mapping, for training and simply because mapping all the nearside of the Moon was such a big job. Bryson was already upset that the maps were behind schedule. In mid-1964 their commute was shortened to a few miles when NASA, under a program funded by William Brunk of the Office of Space Science and Applications (OSSA), built a thirty-two-inch reflector telescope on Anderson Mesa, just south of Flagstaff, dedicated to providing geologic maps of the Moon and staffed by personnel from USGS. David Dodgen and Elliot Morris were the guiding hands while the observatory was under construction, and it later became Elliot's small kingdom, supporting many staffers who spent cold nights at the eyepiece to complete their assigned maps.

Although Bryson had warned me he thought Gene was overloaded with ongoing projects, I intended to offer to support some work at Flagstaff if they could take on additional projects. Our meetings went well, and we agreed to work together on post-Apollo mission planning. The topography and geology of the surrounding area would be ideal for testing some of our ideas on conducting lunar missions with long staytimes, and it was obvious that Gene and his staff passionately wanted to be involved in exploring the Moon. To alleviate Bryson's worries, Gene assured me he could hire extra staff for this new work. We shook hands on developing an interagency funding transfer, and I went back to Washington to start the paperwork. Our handshake would lead to almost $1 million a year in cooperative work, with my office covering all aspects of post-Apollo lunar exploration. By the time the Apollo missions were under way, Shoemaker's team would receive almost $2.5 million a year from NASA to cover its many assignments.

With the paperwork in motion to transfer funding to Flagstaff, Gene began to assemble more staff. He did this with new hires as well as a little Shoemaker "suasion" of USGS personnel at other offices around the country. He had a good nucleus already on site, and to the adventurous recruits this was a mission unparalleled in USGS. A few old hands and a number of younger USGS staff as well as some new hires soon signed up; some reported to the office in Menlo Park, California, to augment the ongoing work there, but most came to Flagstaff. By 1965 Gene had major pieces of many NASA pies: Ranger, Surveyor, Lunar Orbiter, lunar geologic mapping, astronaut training, the job of principal investigator for the first Apollo landing missions, and post-Apollo science planning. At the height of our efforts, in 1968, over 190 USGS staff members and university part-timers were working at several locations in Flagstaff, including offices in a new government complex north of town.

The primary ventures my office funded entailed laying the groundwork to justify the longer-duration post-Apollo missions. This effort soon merged with a need to influence how the Apollo missions themselves would be conducted. With funds beginning to come in from other NASA offices, Gene organized his staff into three offices: Unmanned Lunar Exploration under the direction of John "Jack" McCauley, to cover the ongoing work for Ranger, Surveyor, and Lunar Orbiter; Astrogeologic Studies at Menlo Park under Harold "Hal" Masursky; and Manned Lunar Exploration Studies directed by Don Elston, the last funded primarily by my office.

Our first order of business was to determine what equipment and experiments could or should be included on the post-Apollo missions. We incorporated some of the early results from the MSFC contractor studies as well as the ideas Gene and his staff had begun developing for the Apollo flights. Hand in hand with these studies went the need to define how the astronauts could best accomplish the tasks within the constraints of their space suits and the limitations of their life-support systems. What combination of equipment and procedures would make the most sense from the standpoint of scientific exploration?

In mid-1964 a letter was sent to MSC, over Verne Fryklund's signature, outlining our need for space suits and support technicians to carry out our planned simulations. It requested an inventory of vacuum chambers where we might test the equipment with suited test subjects. We expected that by 1967 we would want to use vacuum chamber tests to demonstrate that, wherever we were in our studies, equipment design, and procedures, the astronauts could carry out the required tasks. Max Faget's response about vacuum chambers was encouraging.[2] Two large, man-rated chambers, A and B (the larger one ninety feet high and fifty-five feet in diameter) were planned for such simulations. He noted that chamber A could sustain tests lasting several weeks, fitting in nicely with our proposed post-Apollo timeline. We thought Max might be having a change of heart about supporting our needs, since the specifications for the chambers came from his office and the only proposal for such long-duration simulations we were aware of came from us. Until this point there had been no exchange of information between the two organizations, so perhaps Max had paid more attention than we thought to Evans's earlier briefing.

The situation on space suits was not so encouraging. Borrowing space suits and technicians for simulations away from MSC would be difficult because both were in short supply. Through the intervention of USGS's Gordon Swann, then stationed at MSC, and others working with the astronauts there, we were able to obtain a surplus Gemini space suit that we trained two staffers at Flagstaff to wear for field simulations. It was not a very satisfactory suit to use in the field, because it was not designed for walking when pressurized, and it was difficult for the wearer to bend at the waist to conduct typical fieldwork. Gemini astronauts either sat in the capsule or, for EVAs, stood almost upright at the end of a tether. But it was useful, especially in the sense that it drove home how difficult it would be for the astronauts, even in a better space suit, to do the equivalent of routine geological fieldwork.

In October 1964 Gordon Swann joined Elston's group, transferring from his work at Houston teaching geology to the astronauts. Gordon brought his insight on how to meet the astronauts' requirements into everything we were doing, based on his day-to-day interactions with them on their training trips. Gordon soon became our primary suited test subject, pouring gallons of sweat into the boots of our borrowed space suits during his many simulations.

As our studies at Flagstaff accelerated, Elston and his staff began to develop several simulation sites nearby. One of these, just east of town, became a convenient place to test our ideas. In July 1964 Bill Henderson and I went to Grumman to have the model shop build a high-fidelity, full-scale replica of the LEM ascent stage as the starting point for our field simulations. The replica was delivered a few months later. We mounted it on a truck bed, and it was carried back and forth to the field when needed.

With additional help from MSC, we soon graduated to a prototype Apollo suit, which made it much easier to conduct realistic fieldwork, since it incorporated a portable life-support system (PLSS) that let us do away with hoses and hand-carried cooling systems. In June 1965 Gordon Swann and Joseph O'Connor were given their first indoctrination into the use of Apollo-type space suits at MSC.[3] From that point on, whenever we could obtain the loan of such a suit, we would rehearse and simulate at Flagstaff all the tasks we were planning for the astronauts.

Our simulations and field tests led to the design of various tools and equipment to ease sample collection and permit the observation and mapping of geological features. Ideas were tried and rejected and equipment was built and discarded as we learned what would work best. For example, during our field simulations, the USGS "astronauts" practiced viewing the surface from the overhead hatch of the LEM mock-up carried on the back of a truck to obtain, more or less, the correct elevation above ground level. Their experience at taking advantage of this high observation point was passed on to the crews and led to David R. Scott's decision on *Apollo 15* to stand in the overhead hatch to plan his surface activities and traverses at the landing site. Dave Dodgen and

Walter Fahey designed and built a LEM periscope like that recommended earlier for the Martin study (with a few more frills), and it was used successfully during some of the simulations to determine how to study a landing site before the astronauts began their EVAs.

At this point in our work Gene had the good fortune and foresight to bring on board a young geologist who had just finished his graduate work at Harvard—Harrison H. "Jack" Schmitt. Jack, full of enthusiasm and energy, soon became a leader in our simulation efforts, and with his firsthand involvement in planning post-Apollo missions at Flagstaff, he began his journey toward becoming (so far) the only professional geologist to walk on the Moon.

We were beginning to make real progress. Not only were we closing in on future tool designs that would work well with a space-suited astronaut, but we were also developing ways for teams back on Earth to process the information that would come back from the Moon in the form of verbal descriptions, experimental data, and perhaps television pictures. At this time a television camera for use on the Moon was not a potential payload item for the Apollo missions. But we believed it would be an invaluable tool for the AES missions, so we usually carried one during our field simulations. We would review the tapes when we returned to the office to complete the analysis of the simulation. We took the next step and set up relay towers on Mount Elden, north of Flagstaff, that let us send the pictures back from the field to an office in the Arizona Bank Building in downtown Flagstaff. After we ironed out the kinks of getting voice and pictures back from the field, we started to design a facility we named Command Data Reception and Analysis (CDRA), where a team of geologists could convert field data in real time into a geologic map. Not only would our planned Moon traverses include geological observations and measurements, but we envisioned collecting geophysical information along the route such as gravity and magnetic field measurements. We knew that AES missions would return so much information, collected during miles of traverses by astronauts riding on some type of vehicle, that it would be essential to process the information in near real time. If we could do this, we believed we could redirect the crews or suggest additional surveys to flesh out the picture we were developing of their landing site.

As our CDRA work progressed we brought our ideas to the attention of MSC. This revelation of how we thought the post-Apollo missions should be conducted stirred up a hornets' nest. We were told in no uncertain terms that the idea would never be approved. Scientists on Earth talking directly to astronauts on the Moon? Scientists second-guessing the astronauts on what to do or how to do it? No way! We were told to cease work along these lines. We chose to ignore this "guidance" and continued to improve our vision of how this could be done.

The ALSS-AES missions permitted longer surface staytimes, but to complete the mission and return home the CSM would have to stay in orbit as long as the astronauts were on the Moon's surface. We began serious study of how we could take advantage of having the CSM in orbit for such a long time. With modifications, in some respects easier to project than extending the LEM staytime, the CSM could remain in orbit for two weeks or longer. What should we do with a CSM that might make three hundred or more orbits of the Moon while the astronauts were on the surface? It seemed obvious: map the Moon from orbit with whatever instruments the CSM could accommodate. In the early stages of these studies we looked at fully automating the CSM sensor package and perhaps converting the LEM to carry three people so that one astronaut would not have to remain alone in orbit on board the CSM but could be on the surface to share the workload. All this appeared possible. We then enlisted the aid of USGS to come up with a conceptual, remote-sensing payload for the CSM. This in turn led to investigating how to tailor the astronauts' surface activities to provide the "ground truth" that would improve the value of the data returned by orbital sensors. The suite of sensors proposed for the CSM included multi-spectral photography as well as spectrochemical, microwave, and radar instruments that would let us extrapolate the data collected at the landing sites to broad regions of the Moon.[4]

By 1965, three years had passed since the last National Academy of Sciences summer study that led to the Sonett Report. In the intervening time we had learned a lot. Careful study of the close-up views of the lunar surface taken by Ranger increased our confidence that "normal" geological and geophysical studies could be planned for the astronauts. The summer of 1965 was selected as the next date for the Academy to review the status of space science, this time at Woods Hole, near Falmouth, Massachusetts. Dick Allenby and I thought this would be a good opportunity to take advantage of the assembled "Academy experts" such as Harry Hess, Aaron Waters, and Hoover Mackin. I hoped to convene a working group similar to Sonett's to review our progress and make some specific recommendations for Apollo and post-Apollo science operations.

We made a few calls to see if some of the invited Academy members would agree to extend their time at Woods Hole. Most agreed to stay—it didn't take much persuasion, since it was such a beautiful spot to be working in the middle of summer. I went to Woods Hole to see if a follow-on meeting could be arranged. In contrast to the twenty participants in the Sonett Ad Hoc Working Group, we envisioned a much larger attendance, probably more than fifty scientists and engineers, including at least one astronaut.

The National Academy of Sciences owned a large mansion directly on the bay at Woods Hole that had been converted to host its many summer conferences. With porches on all four sides of the house and broad, well-kept lawns, it was a beautiful, almost idyllic, site. The views of the bay from the conference room windows made you wonder how participants could concentrate on the business that brought them there. This was my first visit to Woods Hole, and after seeing the mansion I realized that although it could accommodate the small number of scientists usually invited, it would not serve for the much larger meeting we planned.

A few inquiries turned up no suitable building nearby; we needed a small auditorium for general meetings and several rooms where the various scientific disciplines could meet. Driving around Woods Hole and Falmouth, I noticed the Falmouth High School, a perfect location, and on the spur of the moment went in to talk to the principal (I've now forgotten his name). After a brief introduction he gave me a quick tour and said he was willing to ask his school board for permission to host the conference. A few weeks later he called to say it had been approved, and we began the detailed planning for an event that would ultimately involve more than 120 participants.

Developing specific Apollo science guidelines was the first priority of the conference. However, our primary objective for this summer study was to expose the assembled experts to the results of the MSFC contractor studies that we had undertaken for post-Apollo missions. Also, we wanted to show those from the geological community, outside USGS, what we had achieved in more than a year of mission planning and simulation at Flagstaff. During 1964 and 1965 MSC had been steadily adding to its science staff, mostly in the earth sciences, and the frictions I mentioned earlier had been growing. Here was our chance to show them we had received the support of mainstream scientists interested in solving the major lunar problems. Eight of Faget's staffers were invited, led by William Stoney, John Dornbach, and Elbert King. They participated in two of the working groups and also provided technical advice about telemetry and other capabilities that would be needed to support any proposed lunar science ventures.

Two important attendees were Walter Cunningham and Jack Schmitt: Walt was an astronaut, and Jack was an astronaut-to-be. Jack's selection in the first scientist astronaut group had just been announced, and his personal involvement in our Flagstaff work would be an important step in getting the astronauts to accept our ideas on what to do on the Moon and how to do it. Jack would soon be leaving to start one year of flight training; this conference would be his last official duty as a member of USGS. Walt's astronaut group, the third selected, included many who would become well known, such as Buzz Aldrin and Michael Collins. They had all been given specific Apollo system or technology sectors to monitor and become expert in, besides performing their more "mundane duties" of making the transition from military pilot to astronaut. Some had received Gemini mission assignments. Walt's responsibilities included nonflight experiments, so he was our primary contact in the astronaut corps for any questions about the astronauts' performing experiments on the Moon. Other astronauts were given this duty as we approached the Apollo launch dates and the more senior astronauts, such as Cunningham, turned their full attention to preparing for specific Apollo missions.

Having Walt at Woods Hole lent immediacy to our planning. Here was someone who might actually carry out our recommendations. Astronauts' attendance at meetings like ours was always appreciated. Requests for them to appear all over the country flooded into NASA. The demand had become so onerous that Alan Shepard and Donald "Deke" Slayton finally set up a "duty cycle," with each astronaut spending a week or so making public appearances so the others could get their work done. They called this duty being "in the barrel." Some enjoyed the exposure, some hated it, but all tolerated these distractions, knowing that public relations was part of the job. A separate office was established at NASA headquarters to ensure that the proper priorities were recognized when parceling out this valuable resource. Many requests came from members of Congress, and these were usually put at the top of the list. Although most members supported NASA programs, it was to our advantage to keep them all happy, especially at NASA appropriation times. In any case, Walt was an important addition to our conference, and I assume he was happier meeting with us than on some other public relations assignment.

Walt's message to us on the first day of the conference, however, was not encouraging. Influenced in part by his training and by his own study and analysis of the preliminary mission timelines, he warned us not to overburden the astronauts with scientific tasks. Housekeeping chores would demand a large percentage of their time on the lunar surface. Such things as recharging the PLSS, the astronauts' life-support backpack, maintaining work-rest or work-sleep cycles, and monitoring LEM systems—all essential to their safety and health and undertaken in the cramped living space of the LEM—must take priority over science. This was a sobering introduction to lunar science and colored our working groups' deliberations and corridor talk in the days ahead.

Working groups were established in eight scientific disciplines: geology, geophysics, geodesy-cartography, bioscience, geochemistry, particles and fields, lunar atmosphere measurements, and astronomy. Astronomy was added at the eleventh hour in order to review the preliminary findings of our post-Apollo telescope study and to look beyond Apollo to lunar bases when the Moon could become the site of large astronomical observatories. Such installations might include radio telescopes on the farside where they would be shielded from Earth-made noise. At that time there was no intention to include an astronomy experiment on any of the Apollo missions. One of the members of the astronomy panel was Karl Henize, then at Northwestern University but destined to be in the scientist-astronaut class of 1967. The other seven working groups, however, were all tasked to review and recommend experiments and operations for the astronauts to carry out on both Apollo and post-Apollo missions, both for two-week staytimes and for lunar bases. The number of attendees (123) exceeded our initial plans, and to ensure that the post-Apollo discussions would be favorably covered, we loaded the attendance with MSFC and USGS staff who had been participating in our studies.

Each working group submitted a report summarizing the results of its deliberations, and the conference report, compiled by Jay Holmes with the help of many in attendance, was released just before Christmas 1965.[5] It immediately supplanted the Sonett Report as the authoritative reference for Apollo and post-Apollo science planning and, as we had hoped, fully endorsed our approach to the post-Apollo missions. In some cases the working groups went far beyond the concepts we had been studying at MSFC and recommended much more complex experiments than we had considered. For example, we reported on the early results of our studies on a hundred-foot drill, and the geology working group recommended developing a drill capable of taking cores at least three hundred meters below the surface in order to penetrate any ejecta layer and reach solid rock. Those of us who had been working on the drill studies realized that achieving such a depth would be a real challenge, and after the conference we quickly placed a contract with Bendix to take a first look at how it could be done.

The recommendations of the seven working groups for Apollo experiments are too numerous to list here, and many also pertained to post-Apollo exploration, but a few are important in the context of the science payload that was ultimately carried on Apollo. The geology working group listed two primary questions to be answered by the first Apollo landings: What are the composition, structure, and thickness of the Moon's surficial layer? And what are the composition and the origin of the material underlying this layer? Recognizing that time was the most valuable resource in each mission (reinforced by Walt Cunningham's presentation), the group gave a lot of effort to recommending tools and procedures that would permit the astronauts to quickly gather the information needed. Even assuming that all the post-Apollo missions we were planning took place, only a tiny fraction of the Moon would ever be visited and studied. Thus it recommended that manned lunar orbiters be scheduled as early as possible, carrying a suite of instruments to acquire lunarwide mapping and remote sensing information on the Moon's surface composition.

In addition to the geology working group, the geodesy-cartography and geophysics working groups made recommendations dealing with surveying the Moon from lunar orbit. In 1964, under the direction of Peter Badgley, we had begun initial studies of the types of surveys that could be done from an orbiting CSM. We received over one hundred proposals or letters of interest from the scientific community about conducting these investigations, covering all types of surveys from photography to chemical analyses. The Falmouth conference strongly endorsed the need for such investigations.

The deliberations of the geophysics, lunar atmospheres, and particles and fields working groups produced a list of experiments to study the Moon's subsurface as well as phenomena occurring at or near the surface as a result of interactions with the solar wind or cosmic rays. These interactions were of great interest, since it was difficult or impossible to measure them on Earth because of the interference of the Earth's atmosphere and strong magnetic field. For these experiments the Moon could be used as a huge spacecraft floating in free space, on which to mount unique detectors.

The geochemistry-petrology working group also made an important contribution to Apollo science. Only two members of the working group were NASA employees at the time (Paul Lowman was one), but all who participated would later become heavily involved in the program either as NASA managers or as sample-return investigators. The working group concentrated on outlining the procedures NASA should follow in selecting the scientists and organizations that would analyze the samples returned by the astronauts; many of their proposals had just been received. It also recommended sampling procedures and container designs for returning the samples in as near pristine condition as possible. Finally the members turned their attention to the design of the Lunar Sample Receiving Laboratory (later shortened to the Lunar Receiving Laboratory, LRL) where the samples would be quarantined, opened, examined, and sorted for delivery to the laboratories of designated investigators who would then conduct the special analyses they had been selected to do.

Having received the endorsements we were looking for at Falmouth, we charged full speed ahead at Flagstaff to further define potential post-Apollo missions. Based on the emphasis at Falmouth, conserving the astronauts' time became a major objective of our simulations. We also addressed sample return from these longer missions. The weight allowance for return-to-Earth payloads would be restricted, yet the astronauts would undoubtedly collect many samples during their two-week stay. How could they be sure to bring back the most important ones? We proposed a small sample preparation laboratory that they could use while still on the lunar surface, and one was designed by Joe O'Connor, David Dahlem, Gerald Schaber, and Gordon Swann with the help of other USGS staffers. In an undated "Technical Letter" Jerry Schaber reported on the results of one of the field tests, probably conducted sometime in 1966.[6]

The test confirmed that thin sections of the samples for microscopic study could be prepared in this small laboratory, giving the astronauts, who were receiving some rudimentary training in petrography, a first-order idea of what they had collected. (A thin section is made by sawing rock so thinly that light can be transmitted through the slice, telling a trained geologist its mineralogical composition and something of its history.) On the particular test Schaber reported on, they had included a microscope-television system that permitted simultaneous viewing of the thin sections by both the "astronaut" test subjects and geologists back in the CDRA. As Schaber reported, "It became apparent during the test that such remote petrographic techniques could furnish a great quantity of information . . . far more than could possibly be returned to Earth in the present LEM vehicle concept. . . . The test results indicated that the thin section image alone could be interpreted with surprising accuracy by the CDRA personnel." (Perhaps a lesson for future Mars explorers, who will certainly face the same problems we were trying to address—how to get the most information back to Earth with a limited return payload.) Instrumentation that we studied as part of such a small portable laboratory included rock-cutting and thin-sectioning equipment, a petrographic microscope, several types of spectrometers, a gas chromatograph, and an X-ray diffractometer. This concept was presented a year later at the Santa Cruz summer conference, with the recommendation that the images seen in the microscope be beamed back to Earth so that they could be analyzed by experts, thus reducing the time the astronauts spent studying the thin sections.

Our mobility studies at MSFC were providing us with concepts for several types of vehicles that could be carried on the AES missions. In Flagstaff, Rutledge "Putty" Mills, with the help of others, translated these ideas into a working model by modifying a truck chassis to carry two test subjects. Once we had this vehicle, which we named *Explorer,* we planned all our simulations around its use. In 1966 we took delivery of our *Cadillac* lunar rover, a MOLAB (mobile laboratory) working model that MSFC had built by General Motors, Santa Barbara. It was a Cadillac because this MOLAB model cost $600,000 and had a cab so large that two test subjects could live inside and deploy various geophysical equipment as they drove along, without leaving the cab.

When the MOLAB was delivered to Flagstaff, it created quite a stir. It was an ungainly-looking vehicle with four large, tractor-type wheels supporting a fat, cigar-shaped cab with a rather high center of gravity. Shoemaker, watching it being unloaded from the delivery van and thinking ahead to its use in rugged terrain in the field, declared that the NASA-USGS logos painted on the sides would have to be changed. USGS should appear in large letters on the roof, and NASA should be on the bottom. He was sure that during some future field simulation the MOLAB would roll over, and he wanted any assembled reporters to photograph its ignominious fate with the NASA letters showing as the sponsor and USGS safely out of sight. Gene's recommendation was not followed, but his low opinion of the MOLAB test vehicle design was duly reported to MSFC and caused a few red faces. Unfortunately, funding for the AES–lunar base programs was reduced two years after we took delivery of this vehicle, and we had few chances to use it in the field. After a short time it was sent to MSFC, where it was later put on display.

While Gene and his staff were on the front line trying to shape lunar exploration, we were dealing with the USGS management back in Washington in the persons of the USGS chief geologists, first with William Pecora then with his successor Harold "Hal" James. Our relationships were always friendly, but although it was clear that they liked this infusion of new money, they never seemed totally comfortable with the assignment. Exploring the Moon didn't quite fit into the mission of an old-line government agency that had helped open the West a hundred years earlier. This attitude was evident even though at the turn of the century the United States Geological Survey's first chief geologist, Grove K. Gilbert, had been a pioneer in lunar studies.

Pecora and James, at least publicly, were always strong advocates of working with NASA, and on occasion they would be called on to support lunar exploration at congressional hearings or other forums. And certainly the Survey was receiving a lot of favorable publicity from their association with our programs. When the astronauts were covered by the media during geology training trips in some remote corner of the country, there almost always was a USGS staffer identified as lecturing to them. Once the landing missions commenced, USGS contributions became well known, and participants in the field geology experiment were in constant demand to discuss the missions. Even the most hard-hearted manager in Washington must have felt some pride at seeing his agency so prominently featured with the country's new heroes.

Shoemaker was considered a bit of a free spirit within USGS, and all the money he was receiving from NASA, not through his own congressional appropriation channels, was making him rather independent of his Washington superiors. With his successful creation of the Branch of Astrogeology, Gene decided to relinquish his day-to-day management role and once again reorganized by setting up two branches, Astrogeologic Studies under Hal Masursky and Surface Planetary Exploration (SPE) reporting to Alfred H. Chidester. By this time, starting with the first funding transfers in 1961, NASA had transferred almost $14 million to USGS for its various activities, and the action was just beginning to heat up for it to support the Apollo landings. (In all, NASA transferred over $30 million to USGS.)[7]

With the reorganization, in mid-1967 James sent Arnold Brokaw, a manager with no previous experience in lunar studies, to take charge at Flagstaff and make some further management changes. Brokaw's appearance altered the dynamics of our work with SPE, and though we maintained cordial relations with him, we found that the best way to get things done was to work around him and go directly to the staff we had come to know so well over the past three years. The personnel changes made at SPE soon after Brokaw's arrival put our studies in some disarray. Al Chidester, with whom we had cooperated closely, was transferred and no longer had any role in our work. But with the perseverance and cooperation of Gordon Swann and others, we managed to keep things on track, with our eyes focused on the first landing mission and the hoped-for expansion of our ability to conduct exploration in the post-Apollo era.

By the summer of 1967, with the studies at MSFC and USGS described above under way or completed, we had what I considered to be all the key scientific and operational answers needed to justify more extensive exploration and, eventually, lunar bases. We now felt comfortable providing numbers that would help the scientific community accomplish more productive exploration. Science payloads could be at least 2,500 pounds, including a small vehicle, and the radius of operation at the landing site could be up to five miles. Larger payloads might become available as we continued to learn the full potential of the Apollo hardware; we hoped this would lead to MOLAB missions covering much larger areas on the Moon and establishing lunar bases.

We had a lot of new data to share with the scientific community. NASA headquarters had just announced that it would accept proposals for experiments for the Apollo Applications Program (AAP),[8] the new name for the post-Apollo program supplanting Apollo Extension System. AAP missions were advertised to begin in 1971 and would include both manned lunar orbit and landing missions, the latter with surface staytimes up to fourteen days. In Will Foster's office we decided it was time for another summer study to gain more support from scientists for post-Apollo exploration and to encourage them to propose new experiments for the AAP missions. Although the AAP was not yet approved, we thought the announcement was the first step toward its formal recognition, and we wanted to be sure there would be an overwhelming response of new experiments.

Newell and Foster persuaded Wilmot "Bill" Hess, the newly installed head of the Science and Applications Directorate at MSC, to act as the official host of this conference. The idea was to show the scientific community that under his direction MSC had turned over a new leaf and science would now get the attention it deserved in the Apollo program and any programs that might follow. Until Bill's arrival, complaints from lunar scientists had been steadily building, and some MSC offices gave the impression that they knew best what science needed to be done and would do it their way. Don't call us, we'll call you—maybe. MSC was already managing several Apollo science hardware contracts, which added to the concern.

Bill Hess, a physicist, was chief of the Goddard Space Flight Center (GSFC) Theoretical Division when he was asked to transfer to MSC at the end of 1966 to lead a new science directorate. His primary mission at Houston was to reorganize the ongoing science efforts and then evaluate why MSC was held in low esteem by many of the scientists involved in Apollo. A tall, heavy man with a commanding presence, Bill was easygoing but with a touch of steel. He had outstanding scientific credentials and knew NASA politics inside out. We all thought he was the perfect choice for the job. I had come to know him well while he was at GSFC and during the Falmouth summer study, and I knew he would be easy to work with. Perhaps a new day would dawn on our relations with MSC.

Hess had an immediate impact on relations with NASA headquarters. Now, for the first time, we had a senior manager on site who was sympathetic to our concerns and who would return our phone calls, a courtesy seldom extended before his arrival. But he never really became one of the inner circle of MSC managers, and the hoped-for improvements were temporary. One problem was that although he was starting a new directorate, he inherited some of the people from Faget's office who had been giving us all such a hard time—it isn't easy to fire or transfer civil servants. In his two short years the climate for science improved, but this was soon reversed by his successor.

The site selected for the 1967 conference was the new University of California campus at Santa Cruz. Aaron Waters, a noted geologist and coinvestigator on Shoemaker's Apollo Field Geology Team, had just joined the staff at Santa Cruz and served as the unofficial host. Over 150 people joined us at Santa Cruz, representing all the geoscience disciplines and including a few astronomers.[9] MSFC sent only two observers to the conference, because by this time the decision had been made to manage all Apollo science at MSC, and MSFC quickly phased out of most lunar science studies. Goddard Space Flight Center was well represented, led by Isadore "Izzy" Adler and by Jack Trombka, who had returned to GSFC after his stint at headquarters. They wanted to map the lunar surface extensively from orbit using newly developed sensors. Thirty MSC staffers from various organizations attended, including Faget himself, as well as three astronauts: Deke Slayton, Jack Schmitt, and Curtis Michel (a member of Jack's 1965 scientist-astronaut class).

The large number of MSC attendees attested to Hess's new influence and perhaps to the recognition that these summer studies were important in shaping lunar science. They came prepared to push their point of view on what science the astronauts should conduct and how it should be done. (I should clarify my criticism of MSC, since it does not apply to the organization as a whole. At this time we were able to work with the MSC science staff, although with difficulty, and Hess's interest in changing the working relationships with headquarters and the science community was smoothing some of the rough edges. Our relations with other organizations at MSC were usually good, and when I was in Houston I could confide in many friends at MSC or sit down at dinner and discuss the state of NASA.)

As we did at Falmouth, we asked the attendees to think in terms of grand exploration missions, and we had the numbers to allow this. With the newly named Apollo Applications Program would come one of the last attempts at persuading Congress and the administration to continue exploring the Moon after the initial Apollo landings. We hoped that the Santa Cruz conference would stimulate the scientific community to continue supporting lunar exploration in spite of growing frustrations with attempting to influence the scientific content of Apollo.

Our daily sessions were divided into eight working groups, which reported on their findings at the end of the conference. I attended as secretary of the geology working group, which was led by Gene Shoemaker and Al Chidester (one of Al's last duties before his transfer) and was dominated by USGS staff and university professors who supported the work we had been conducting at Flagstaff. Major recommendations coming out of this working group included (1) increasing the astronauts' radius of operation beyond walking range, estimated to be five hundred feet, by providing wheeled and flying units; (2) developing a dual-launch capability as soon as possible; (3) creating a sample return payload of four hundred pounds; (4) making the geophysical station flexible so we could react to new opportunities; (5) providing an early manned lunar orbital flight to further map the lunar surface in the visible part of the electromagnetic spectrum and other parts as well; and (6) sequencing orbiter and landing site missions that would include landings at the craters Copernicus and Aristarchus. In general, all the recommendations supported the post-Apollo planning we had undertaken in the past four years.

One of the conference's recommendations was of special interest to me and others. A second scientist-astronaut selection was under way at the time of the conference, and I was in the final group under consideration. Knowing of the sensitive nature of crew selection and the competition for slots on the landing missions, the working groups tried to be diplomatic when making their recommendations for crew training and selection. Also, we hoped that Jack Schmitt would be selected for an early lunar mission, and we did not want to jeopardize his chances by being too aggressive in our advice. The recommendation on astronaut selection and crew training included these words: "For some of the complicated scientific missions in the later part of the AAP, the Santa Cruz Conference considers that the knowledge and experience of an astronaut who is also a professional field geologist is essential." At the time I hoped they would be to my own benefit during the selection of the next class of scientist-astronauts.

Although the Santa Cruz conference endorsed the need for missions after the scheduled Apollo flights, time was running out for AAP.[10] The Santa Cruz attendees, representing many renowned scientists, had proposed important studies on the Moon that were not planned for Apollo. These experiments would require payloads and resources beyond what was anticipated for the Apollo flights. By the time the conference came to a close we knew that NASA budget submittals for fiscal year 1969 would not include funds for missions beyond the already funded Apollo flights. What exquisite timing.

At this point in my government career I had seldom come into contact with the Bureau of the Budget (later named Office of Management and Budget), but in the ensuing years, as a senior official at several agencies, I would frequently meet and argue with its staff members. The original "faceless bureaucrats," they had enormous authority and no responsibility. If a program failed or struggled because of arbitrary funding cuts, the agency and program managers would bear the brunt of the failure, not the BOB/OMB staff members who had wielded their red pencils. I don't recall ever encountering an OMB staffer who had managed a real program; they were blissfully unaware of program complexities other than dollars. In spite of this rejection by BOB, we continued to plan for dual-launch missions and extended lunar surface staytimes. We could always hope that the upcoming election might produce an administration more friendly to lunar exploration.

In the fall following the Santa Cruz conference, some major organizational changes took place at NASA headquarters that altered the nature of planning for both the Apollo missions and the missions that might follow the first Apollo landings. With these changes several of us, from various offices, moved to the Apollo Program Office. But before continuing the story of Apollo and post-Apollo science, let's turn back the calendar to where we left Apollo science at the end of chapter 1.

Science Payloads for Apollo:
The Struggle Begins

In July 1960, before President Kennedy's dramatic declaration that we would send men to the Moon and return them safely and before Alan Shepard's successful Mercury launch, NASA announced that it was considering manned circumlunar flights. This unnamed program proceeded slowly, responding in some degree to what the Soviet Union was accomplishing. Then, pushed by growing concerns about Soviet success in space and relying on NASA managers' assurances that a manned lunar landing was achievable, the president made his historic national commitment, soon endorsed by Congress.

Little by little, with many twists and turns along the way, the program matured. It was given the name Apollo, and its "mission architecture" was agreed to. Mission architecture comprises those aspects of a typical mission (size of the rocket stages, spacecraft design, flight trajectories, timelines, etc.) required to accomplish its objectives. This "architecture" would eventually control or shape the scientific experiments the Apollo astronauts would conduct. Here I discuss these aspects of Apollo and briefly describe the supporting programs, both manned and unmanned, that Apollo science depended on. Then later in this chapter and in the following ones I tell about the struggle to add science payloads to the missions. To maintain the continuity of particular topics, I sometimes depart from a strict chronological sequence.

After the lunar orbit rendezvous (LOR) approach described in the introduction was adopted, work began to build the *Saturn V* launch vehicle and two spacecraft: the three-man command and service module (CSM) and the lunar module (LM; earlier called the LEM, lunar excursion module). Lunar missions utilizing LOR required the *Saturn V* to first place the spacecraft in Earth orbit and then send them on to lunar orbit. After doing their jobs, the initial two stages of the *Saturn V,* the S-IC and S-II stages, would be jettisoned, reenter the Earth's atmosphere, and burn up. The upper stage, the SIVB, with the CSM and LM spacecraft attached, would then be sent to the Moon or, in NASAese, put into a translunar injection. Once safely on the way and coasting toward the Moon, the CSM would separate from the SIVB, turn, and pluck the LM from the SIVB, where it had been stored just behind the CSM inside a protective fairing. The SIVB stage, with no further function and essentially depleted of fuel, would go its separate way, deliberately steered away from the Moon in the first flights to avoid any interference with the mission. Together the CSM and LM would continue on to the Moon. Upon arrival the spacecraft would use the CSM engines to brake into a low lunar orbit.

Once in lunar orbit and after all systems had been checked, two astronauts would enter the LM, separate from the CSM, and descend to the lunar surface, leaving the third astronaut in lunar orbit in the CSM to await their return. The LM would be a sophisticated two-stage spacecraft comprising the descent stage that fueled the landing maneuvers and the ascent stage in which the astronauts would travel to the Moon's surface and return to rendezvous with the CSM in lunar orbit. If the landing had to be aborted, the LM descent and ascent stages could separate while in flight and allow the astronauts to rendezvous with the CSM. The LM also included the small cabin in which they would live during their stay on the lunar surface. The two stages would carry the equipment for use on the lunar surface. After leaving the Moon and meeting the CSM in lunar orbit, the ascent stage would be jettisoned, and when its orbit decayed it would crash on the Moon.

Similarly, the CSM was a multifunction spacecraft. As the name indicated, it had a dual purpose, serving as a command ship and a service module. The command module portion was the control center of the spacecraft and the astronauts' home on both the voyage to the Moon and the return to Earth. The command module pilot would monitor the other astronauts' progress on the lunar surface and, on later missions, conduct sophisticated experiments. After the astronauts left the Moon's surface in the LM ascent stage and achieved a lunar orbit, it was the CSM pilot's job to rendezvous and dock with the LM ascent stage so the astronauts could transfer to the CSM along with any material they brought back from the lunar surface. The rear end of the CSM, the service module, was primarily a rocket and logistics carrier. It supplied power and life-support expendables for the command module and propulsion to permit a wide range of maneuvers. Most important, it provided the propulsion to take the CSM out of lunar orbit and bring the astronauts home. Once Earth reentry was ensured, the service module would be jettisoned. The command module would reenter and parachute to an ocean landing.

With this abbreviated description of the Apollo hardware as background, I can begin to tell how we struggled to place science payloads on board Apollo. Because the *Saturn V* had to lift some six million pounds of equipment and fuel from the Earth's surface to Earth orbit and the succeeding stages had to perform efficiently in order to send as large a payload as possible to the Moon (much of it in the form of rocket fuel), the weight of the total *Saturn V* and all the many components rapidly became an overriding design concern. On my first visit to Grumman in 1965, at Bethpage on Long Island, to see an early version of the LEM, weight concerns were high on the agenda. After a brief walk around this peculiar contraption with long spindly legs and tiny triangular windows, we attended a status review. The LEM was in trouble; among the issues covered was how to reduce its weight. If this could not be done, the problem would affect all the Apollo systems and subsystems. The Grumman engineers took this so seriously that they were counting rivets as they modified the design to achieve their weight targets. And here we were, trying to convince management to add hundreds of pounds of science payload to the LEM; without question it would be difficult.

Based on the scientific guidelines mentioned in chapter 1 and on the Sonett Report, in November 1963 I made a quick parametric study to determine what science might be done at any point in a typical Apollo mission, from translunar injection to the final return to Earth.[1] This brief analysis focused primarily on the "what-ifs": for example, what if the first astronauts achieved lunar orbit but could not descend to the surface; what if they descended to the surface but couldn't land; and what if they landed but couldn't exit the LEM? My purpose was to identify instruments and equipment that would be needed to make the most of each opportunity and set priorities for what should be included in the (probably small) science payload. As one might guess from the list of what-ifs, a camera, or several cameras, would have high priority. The Martin Marietta contract discussed in chapter 3 was a direct outgrowth of this analysis, concentrating on what to do if the astronauts made a successful landing but were not permitted to leave the LEM.

Two months later, in February 1964, after our office further reviewed the Sonett Report and the Apollo science program guidelines, Will Foster sent the Space Science Steering Committee of the Office of Space Science and Applications a memorandum providing a preliminary listing of the scientific investigations that should be considered for Apollo.[2] This memo, which I discuss in detail in the next chapters, defined the areas of interest for each scientific discipline and listed the scientists who would be asked to help plan individual experiments. With this additional guidance, Ed Davin, Paul Lowman, and I did a more careful analysis of the what-ifs and wrote a short report in early June outlining a program of Apollo scientific investigations covering the first seven Apollo landings, the approved program at that date.[3] We went into some detail for the first landing mission, assuming it would allow only four hours of extravehicular activity (EVA) on the lunar surface. We also described a "limited mission profile" that permitted only one hour of EVA. Both the one-hour and four-hour EVA plans took into account our limited knowledge of the constraints that might be in effect based on prototype Apollo space suits. A primary reason for our report was to have a handout reflecting Manned Space Science's position available for distribution at the Manned Spacecraft Center Lunar Exploration Symposium that was scheduled for June 15 and 16, 1964.

At the symposium we and many of the scientists named in Foster's memo were exposed to MSC's view of what could be done on the lunar surface, allowing for probable operational constraints. Lively debates took place, with the science side attempting to understand and relax these constraints so that more scientific work could be accomplished. The science planning team members described the experiments they hoped to have the astronauts deploy and the types of studies and observations that would be needed. Everyone left with a much better understanding of what lay ahead before we could all agree on the best methods of exploration during the missions.

The symposium led us to rethink several of the what-ifs. In particular, what if the astronauts could not leave the LEM to deploy the experiments they were carrying? Members of the seismology panel thought the seismometer could be designed to be turned on from Earth while still in the descent stage equipment bay, thus allowing some readings of the Moon's seismicity, especially if any large natural events occurred near the landing site. MSC had pointed out that the landings would take place at low sun angles and there was a fifty-fifty chance that after touchdown the LEM windows would be facing the Sun, making photography from inside the LEM difficult. If the astronauts could not leave the LEM, the landing site would be poorly documented. We again suggested adapting the LEM telescope or adding a periscope to permit photographs, but we received no encouragement.

Another interesting discussion dealt with speeding up one of the housekeeping tasks—recharging the space suits' life-support batteries. In the preliminary timeline that was presented, six hours were allocated for the recharge while the astronauts were back in the LEM, thus restricting the total EVA time. The Crew Systems Division pointed out that simply swapping out new batteries could reduce this time to fifteen minutes, and the spent batteries could be recharged during any subsequent downtime. Our office proposed reserving some of the science payload for additional batteries (about five pounds each). We updated our June report to reflect our new knowledge.[4] Fortunately, payload weight allowances grew and we were spared a painful trade-off, giving up science payload for additional batteries to get more EVA time.

During the symposium two trends were becoming evident. We were more and more at odds with the MSC Engineering and Development Directorate on how to incorporate science on the missions and even on what experiments should be carried. Yet we were developing a close relationship with members of the Crew Systems Division, which had day-to-day contact with the astronauts in developing operational protocols covering not only future scientific work but all the astronauts' other activities. Like our good working relationships with other MSC offices, this one would prove invaluable in the years ahead, since they would act as intermediaries with MSC management.

Three other programs—Ranger, Surveyor, and Lunar Orbiter—were also under way at this time, designed to support the manned lunar landings. These were unmanned programs managed by OSSA at NASA headquarters and implemented by NASA field centers: the Jet Propulsion Laboratory (JPL) for Ranger and Surveyor and Langley Research Center for Lunar Orbiter. Both the Ranger and Surveyor projects were initiated in the late 1950s, not to support Apollo but as purely unmanned scientific programs. However, these two projects soon succumbed to the needs of the larger Apollo program. Eventually both were reduced from their original scope, reflecting both funding and priority concerns, but their primary functions endured. Ranger would provide early detailed pictures of the lunar surface, so necessary in planning for the manned landings, and Surveyor would demonstrate the ability to soft land a spacecraft and would also send back some close-up pictures of the lunar surface and engineering data on its characteristics. Lunar Orbiter had the specific objective of taking detailed photos of potential Apollo landing sites.

The programs would be increasingly complex, testing our ability to operate spacecraft at lunar distances, which could not be done in the late 1950s when Ranger and Surveyor were conceived. Among other considerations, a network of communication stations would have to be built around the world to permit round-the-clock tracking and control of the spacecraft. The three projects represented important technological advances, but they would be far less difficult to develop and operate than the Apollo missions. By 1963 the Soviets had already sent six partially successful Lunik spacecraft to the Moon; with these and their manned Earth orbital flights, they were considered far ahead of us in developing and operating such complicated missions.

Leading up to the Apollo flights, the Mercury and Gemini projects made NASA confident that it had conquered the hazards of manned space flight. *Faith 7*, piloted by Gordon Cooper, the last spacecraft in the Mercury program, had already splashed down in the Pacific by the time I joined NASA. The six manned Mercury flights accomplished all the goals assigned to the project and more. NASA had graduated to the next big step—Gemini—with new confidence in its ability to safely launch men and equipment into space and recover them at sea even if the splashdown occurred far from the planned recovery point, as on Scott Carpenter's *Aurora 7* flight. Apollo would also be designed around an ocean recovery, the final act in each mission. The Soviets' manned program made all its recoveries on land, usually somewhere in one of the eastern republics. Ocean recovery was viewed as less risky in case of reentry problems, and with our large naval forces deployed around the world, ocean recovery of any Apollo crew was judged easier.

When I joined NASA in late 1963, all the Gemini flights still lay ahead. They were designed to provide the training for the more complex space operations needed for the Apollo missions. The Gemini spacecraft carried two astronauts in cramped quarters. They would perform maneuvers never before attempted in space, such as a rendezvous with another spacecraft and the movements outside the Gemini capsule that NASA called extravehicular activity and the press dubbed space walks. Considering that men had been operating in space only four short years before the first manned Gemini flight, these missions would be truly groundbreaking. The Soviets were still ahead in number of missions and man-hours in orbit, but their spacecraft were not capable of maneuvering like the Gemini spacecraft, and their EVAs had been short, tethered stunts. On the Gemini EVAs the astronauts would perform specific tasks like those that might be needed on an Apollo mission.

Like the Mercury program, Gemini accomplished all its planned objectives. *Gemini 8* was especially memorable for me. It was launched on March 16, 1966, its crew consisting of Neil Armstrong and David Scott. The launch coincided with one of the aerospace industry's most important social events, the Goddard Memorial Dinner in Washington, D.C. In 1966 this dinner attracted aerospace luminaries from both industry and government. The Goddard trophy, named after Robert Goddard, the father of United States rocketry, was awarded to an individual or group in industry or government chosen for special contributions in advancing the space program during the past year. The award on this night went to President Lyndon Johnson, with Vice President Hubert Humphrey accepting for the president.

In 1966 the Goddard dinner was a rather intimate gathering of some three to four hundred guests. I say intimate because today the dinner attracts more than two thousand, with the men in black tie or dress uniforms and the ladies in formal gowns. The 1966 dinner, as I recall, had few women, and all the civilians wore business suits. Government attendees were usually the guests of some company, and the invitations were—and still are—carefully orchestrated to avoid any perception of conflict of interest, although it was clear who your host was. Tickets cost about $35 in those days; today they are $175, not an insignificant sum then or now. I was the guest of Bendix, one of the contractors working on the studies I was sponsoring at Marshall Space Flight Center.

As the guests at the head table were being acknowledged, including the vice president, there was an interruption in the speeches. Someone walked up and whispered in George Mueller's ear. He nodded and said a few words to several other NASA managers seated near him, then they all got up and filed out. The room buzzed, but the program continued with the vice president's speech accepting the prestigious award on behalf of the president. It was several hours before any of us knew why Mueller and the others left. *Gemini 8* had experienced a serious problem.

In the first scheduled space docking between a Gemini capsule and an earlier-launched Agena target vehicle, the two spacecraft, after being joined for about thirty minutes, began to spin rapidly, forcing Armstrong to back away.

One of the capsule's thrusters had stuck open, causing the rapid rotation; only through Armstrong's extraordinary skill were they able to bring the spacecraft under control. This complication forced an early termination of the mission, and not all its objectives were achieved. But Armstrong's and Scott's cool behavior in this dangerous incident (some estimated they only had a few more seconds to correct the problem before centrifugal force would have caused them to black out) undoubtedly elevated their position in the astronaut corps and put them on Deke Slayton's short list of prime candidates for the later Moon landings.

In early 1964, with the ink barely dry on his agreement to coordinate science activities between OSSA and the Office of Manned Space Flight through Will Foster's office, Mueller took the next step toward controlling what science would be carried out on the Apollo flights. Many types of experiments besides those falling under OSSA's purview were being suggested by other offices. Some dealt with the life sciences, primarily advocated by MSC's Medical Directorate, and a series of engineering experiments were being proposed by several NASA offices as well as the Department of Defense. To establish uniform requirements for the experiments and set priorities for inclusion on the flights, Mueller established the Manned Space Flight Experiments Board, with membership from all the competing offices but chaired by OMSF.

Attention to science concerns was advancing on another front at MSC. In 1963 Max Faget had established a new division in his Engineering and Development Directorate, called Space Environment, that would interact with the scientific community. At the beginning of 1964 this new office, led at first by Faget, began to address two important questions: How would the returned samples be handled, and who would be responsible for receiving, cataloging, archiving, and distributing samples to those approved to do the analyses? MSC, led by Elbert A. King, a recently hired geologist, began lobbying to build a small laboratory to carry out these tasks. At the end of 1964 Homer Newell asked the National Academy of Sciences' Space Science Board to determine if there was a requirement for a special facility to handle the samples. The board, chaired by Harry Hess, forwarded its report in February 1965.[5] It endorsed the need for a rather modest laboratory that, among its other functions, would quarantine the lunar samples for some unspecified time to ensure that they did not contain dangerous pathogens. With the release of the report, a major difference of opinion surfaced between headquarters and MSC on where the lab should be.

The report pointed out some of the pros and cons of establishing such a facility at MSC but noted that the committee did not believe it should be there. Those of us in Foster's office who had an interest in the outcome of this debate were dead set against the lab's being built at MSC. Based on our earlier attempts to work with some of the MSC science staff and with particular individuals in the Space Environment Division, we were suspicious that their wanting to build a special sample facility at MSC was a devious attempt to control all the returned samples and thus justify having MSC staff carry out most of the analyses. We advocated considering an existing laboratory such as Fort Dietrick in nearby Maryland, which already had experience in handling dangerous biological material, as the repository for the samples.

Congress also became involved, since a new facility would be costly. In spite of all these objections, the Lunar Receiving Laboratory was built at MSC, and King was later named the first curator. Although some of our fears were realized in the ensuing years, the LRL was very successful. A major reason our office accepted MSC as the LRL location was the appointment of Bill Hess, whom we all trusted to make the right decisions on how it would operate. Hess oversaw staffing and the development of procedures that would ensure the integrity of sample analysis and control sample distribution.

The many functions the LRL would perform required a unique design. Because of its extraordinary mission and the controversy over its siting, during the next several years I watched the construction with interest on my many visits to MSC. One of the concerns the National Academy of Sciences committee had about locating the lab at MSC was the construction of a radiation-counting facility. It had to be built far below the surface (fifty feet) to shield selected samples from background radiation. Gamma radioactivity had to be measured as soon as possible after the samples arrived, before the shorter-lived nuclides decayed. These sensitive measurements (never before attempted on such fresh extraterrestrial material as the Apollo samples would represent) would furnish information on the origin and history of the samples and of the Moon itself. During counting and storage, the samples would have to be held in a room that was not only below ground but heavily encased in steel plating and other types of shielding. It was feared that underground construction at MSC, where the water table was high, would greatly increase the cost of the lab. I attended the unveiling of the low-level counting facility and heard about how difficult it had been to find steel for the outer shell that would meet the stringent low-radiation standards. Steel cast after the United States and Soviet nuclear tests would be contaminated by the fallout from these tests so that background radiation would be too high even with a thick layer of dunite between the outer shell and the counting laboratory itself. The contractor finally found some scrap steel from the hull of a ship built before World War II.

In addition to the low-level counting facility, the LRL had several other unique features, including crew quarantine living quarters. After splashdown and before leaving the CSM, the astronauts would don special isolation garments so as not to come into direct contact with the helicopter recovery team that picked them up and flew them to the carrier. Once on board the carrier the astronauts would be rushed to the mobile quarantine facility, which looked suspiciously like an Airstream trailer without wheels (it was built by Airstream to NASA specifications). You may have seen pictures of the *Apollo 11* astronauts at a window in the MQF, waving to President Nixon on board the carrier USS *Hornet*. The MQF was designed to be airlifted back to Ellington Air Force Base, then it would be trucked to MSC and the LRL. Once at the LRL, the astronauts and the physicians who had volunteered to accompany them would leave the MQF and pass through an airlock into their quarantine quarters, called the crew reception area, where they would stay for the rest of their twenty-one-day quarantine period. The CM would also be flown back to the LRL, since its interior would be considered contaminated from lunar dust adhering to the astronauts' space suits.

The LRL interior was maintained at negative atmospheric pressure to prevent the escape of any dangerous organisms. When you visited, either to attend astronaut debriefings or to observe sample preparation, you passed through an airlock, popped your ears, and went on about your business. Inside the LRL were a number of gas-tight glove cabinets and vacuum chambers where technicians would open the sample bags, record their contents, and prepare the samples for shipment to the sample analysis principal investigators (PIs) at the end of the quarantine period. The LRL functioned with few problems over the next five years, and it exists today as a curatorial facility, although most of the samples from all the missions have been transferred to another location. Only small amounts of sample material were distributed and analyzed in great detail. NASA still entertains proposals to examine samples from those qualified to conduct some unique study.

Backtracking slightly, in January 1965, over the signatures of George Mueller and Apollo program director Sam Phillips, OMSF issued the *Apollo Program Development Plan*.[6] Originally a classified document (I assume to keep the Soviets from knowing our schedules and other details), the plan was designed to "clearly identify the program requirements, responsibilities, tasks, resources, and time phasing of the major actions required to accomplish the Apollo Program." Consisting of 220 pages of detailed guidance on all aspects of the program, it stated in the introduction that the manned lunar flights would conduct scientific experiments in cislunar space and that the manned lunar landings would be made "to explore the moon's surface and to conduct scientific experiments." All the various parts of the program were identified from the development of the *Saturn V* and its several components to the launch facilities and ground tracking stations. The plan also identified which NASA center or other government agency would develop each of the pieces. Despite Mueller's and Newell's recent coordination in establishing the Manned Space Science office, the plan is remarkably silent on how scientific undertakings would be managed or who would ensure that experiments would be ready when needed. Reading between the lines, you could assume that MSC had this assignment under the heading of Flight Mission Operations, but scientific operations were not specifically called out. The Manned Space Science office receives one mention, as a title only, in a facilities analysis matrix. Why it was placed in that section of the plan is a mystery—probably an afterthought by the authors. In early 1965 Apollo's objective clearly was to land men on the Moon and return them safely, the few words in this new plan dealing with science notwithstanding.

In 1965 Mueller also established the Apollo Site Selection Board (ASSB). In the beginning the board was chaired by Sam Phillips and included members from headquarters and center offices. Its initial function was to set priorities for Lunar Orbiter photographic coverage to ensure that the pictures needed for selecting Apollo landing sites were adequately identified and scheduled. After Lunar Orbiter successfully completed its objectives, the ASSB turned its attention to the more difficult task of choosing the first and subsequent Apollo landing sites.

In most respects the first landing sites were easier to select than the later sites. The "Apollo zone of interest" was quickly established based on the predicted performance of the *Saturn V* and the Apollo spacecraft. The "zone," bounded by the lunar coordinates five degrees north and south latitude and forty-five degrees east and west longitude, covered—as far as we could tell from telescopic photography—mostly smooth lunar mare areas, another requirement for the first landing. Conditions for touchdown required that the LM come to rest at an angle no greater than twelve degrees from the horizontal, to avoid problems when the ascent stage lifted off. Since one of the LM's landing struts might end up in a depression or the lunar surface might have a low bearing strength, the ASSB was hoping to find areas rivaling a billiard table.

After the initial landing conditions were met, it was anyone's guess where the next landings would take place. Again, overall system performance dictated mission safety rules, which in turn would restrict site accessibility. MSC wanted to stay close to the lunar equator for flexibility. Those of us pushing lunar science wanted to stretch system performance to its limits and land near a variety of important features that promised to answer important scientific questions. Such features usually augured rough landing sites.

While all these assignments were under way, Homer Newell was putting procedures in place that would give OSSA greater influence concerning the experiments carried on Apollo. In addition to the National Academy of Sciences' Space Science Board—a powerful voice for science from outside the halls of NASA that gave him overall recommendations and direction—Newell looked to the Space Science Steering Committee (SSSC) to help oversee the selection of experiments for both the manned and unmanned programs. This committee, composed of government employees, was assisted by several subcommittees that included members from both inside and outside NASA. The subcommittee that dealt most directly with lunar science was the Planetology Subcommittee, chaired by Urner Liddell. It met frequently to review and approve scientific proposals for the unmanned programs, and in 1964 it began to provide OSSA with Apollo science oversight.

Liddell was a strong proponent of unmanned space science and a confirmed skeptic about the value of having man (astronauts) in the loop. His leadership of this subcommittee would create some friction between OMSF and OSSA in the next few years. Liddell had a voice in choosing members, and he selected prominent scientists who supported his low opinion of manned science. Fortunately there was one strong defender of manned science on the subcommittee—Harry Hess, who also chaired the Space Science Board. Hess, a renowned geologist and a professor at Princeton, would soon become one of our leading champions, countering the scientific elite who shared Liddell's opinion that no good science would be accomplished on the Apollo missions. Dick Allenby also served on the subcommittee. He represented our positions on manned science but usually found himself overruled by his former boss, Liddell.

Bob Fudali, never one to mince words, wrote: "The character of Urner Liddell continues to fascinate me. It was most instructive to watch him squelch the junior subcommittee members with his overbearing mannerisms."[7] The Planetology Subcommittee meeting of January 1965 that Fudali was reporting on introduced two new members: Donald Wise, from Franklin and Marshall University, and George Field, from Princeton. Wise later had a prominent role in Apollo science. Since they were the two most junior members, they were undoubtedly the unnamed squelchees.

The agenda for that meeting was long and included discussions of the design and location of the LRL and developments in the "Moon Blink" project. Those attending were asked to rank four experiments proposed for the first Apollo landing: passive seismometer, gravimeter, magnetometer, and micrometeorite detector. The first three experiments did not yet have identified PIs, and the last one was proposed by MSC. The seismometer and gravimeter were given top priority, and a decision on the magnetometer was deferred. The micrometeorite experiment was given the lowest priority as "not germane" to lunar science. MSC sent John "Jack" Eggleston to the meeting to participate in the experiment and LRL discussions. While defending MSC as the future LRL location, he made an interesting disclaimer. In reaction to negative comments from the subcommittee members, Fudali reports, Eggleston said he realized MSC lacked qualified scientific personnel and that it would hire only enough technicians and junior scientists to assist the sample investigators chosen by the scientific community. But MSC soon went back on this pledge and hired a large scientific staff, assigned to Faget's organization. Most would be transferred to the Science Directorate when it was formed, reporting to Bill Hess.

With minimum fanfare, we brought into the program prominent scientists who would develop specific experiments. By this time a good consensus existed on the important experiments to conduct during the Apollo missions. This made it a relatively straightforward task for the Planetology Subcommittee and its parent body, the SSSC, to select PIs. The only potential difficulty would be choosing between well-known PIs wanting to do the same experiment. This competition never arose because the major experiments were proposed by teams of scientists that included some of the most recognized names in their disciplines. The first PI selected under this procedure to lead the Field Geology

Team was Gene Shoemaker. PIs were soon named for all the high-priority experiments.

In June 1965, under the auspices of OSSA, we circulated within NASA the first comprehensive report on the exploration and utilization of the Moon. The report included important contributions from many OSSA offices, since it covered plans for both manned and unmanned lunar exploration extending to 1979.[8] Will Foster's office took the lead in summarizing our current thinking on manned missions, beginning with the first Apollo landing, shown as occurring at the end of 1969 and progressing through dual-launch Apollo Extension System manned orbital and surface missions to the first lunar bases.

We explained the rationale for this mission progression by tying it to the important scientific questions and operations that would justify a continuing program. Many of the studies we had initiated at MSFC were cited to provide the detail the plan required to justify the types of missions referred to in the plan's ninety-six pages. The report concluded by stating, "The lunar exploration program is an important part of the nation's space program. Scientific investigations in this field are a significant aspect of the overall endeavor to advance our capability and to continue U.S. leadership in the adventure into space." Those of us who had been working on manned lunar exploration saw this statement as OSSA's first acknowledgment of the importance of manned exploration. Up to this point we had always felt that the science side of NASA was merely tolerating manned missions while its eyes were on bigger targets— unmanned explorations of the planets.

Just before the Falmouth conference, OMSF published the first *Apollo Experiments Guide,* intended to supplement the announcements of flight opportunities (AFOs) then in circulation or any that might be released by NASA offices about opportunities to carry out experiments on the Apollo missions.[9] A short preliminary guide had been issued in June 1964, peppered with such warnings as "best estimate," "experiments shall be conducted on a non-interference basis," and "specific weight assignments . . . cannot be stated for each flight at this time," to indicate the uncertainty associated with putting experiments on the Apollo missions.[10] The 1965 edition contained more information but continued to demonstrate OMSF's ambivalence about encouraging scientific experiments on the Apollo flights. Eighteen months earlier we had issued preliminary guidelines for Apollo science including a designation of 250 pounds for science payloads. The new guide seemed to be a step backward. It estimated seventeen cubic feet of stowage on the LM and the capacity to return eighty pounds of samples from the lunar surface, but it listed no overall allocation of payload weight on what were termed the early developmental missions. One could interpret the guide to mean that the stowage space might be empty on these flights and that the only "science" conducted would be the astronauts' collecting samples with their gloved hands.

The 1965 guide stated that the Manned Space Flight Experiments Board (MSFEB) would approve the experiments to be carried and outlined the procedures it would follow. The board, nominally chaired by George Mueller but often led by a deputy, consisted of senior managers from headquarters and field centers and one representative of the Air Force Systems Command. Will Foster was our representative for lunar exploration. Experiments would be selected by various NASA offices such as OSSA and then passed to the MSFEB. Those of us who had been trying to increase the science payload allocation looked with deep suspicion on this board because it included members from NASA offices of Space Medicine and Advanced Research and Technology as well as MSC's director, Bob Gilruth. We knew that these offices and MSC had already proposed some Apollo experiments (such as the micrometeorite detector). We could see the limited science payload, however much it ultimately turned out to be, being slowly eaten up and given to what we felt were peripheral experiments, not designed to study the Moon as a planetary body. In later years, when the actual experiments were approved by the MSFEB, Ernst Stuhlinger often represented Wernher von Braun and MSFC, giving us another voice on the board who fully understood what the science community was trying to accomplish for lunar exploration.

As the final filter, the MSFEB would carry out another important function. For all space missions, manned or unmanned, AFOs would usually give experimenters broad guidelines on integrating experiments with the spacecraft they would fly on. But at this early date, 1965, no *Saturn V* boosters or Apollo spacecraft had flown, so many of the integration specifications were guesstimates. Experiment design considerations dealing with such aspects as vibration levels, acceleration forces, shock, and acoustical levels would not be known for some time. In addition, other concerns such as avoiding materials that might cause adverse reactions like electrolytic corrosion or electromagnetic interference (airplane passengers must turn off electronic equipment during the early and final stages of a flight) and a host of other dangerous interactions with the spacecraft or booster could not be completely defined. The MSFEB would be the ultimate judge of whether the experiment, in many cases conceived and designed before final specifications were available, passed the rigid integration criteria and would be approved, rejected, or sent back for modification. Integration of the experiments was a difficult hurdle because experiments also had to pass "astronaut integration" if they required any input from the astronauts, a developing art in 1965. Principal investigators soon learned that if they wanted to participate they needed patience and perseverance and that they must overlook what seemed like strange, bureaucratic rules.

Time was also becoming a factor in selecting and building the experiments. The guide advertised 1968 to 1969 as the need date for delivering the experiments to Kennedy Space Center (KSC). Along with the uncertainties mentioned above, a tight schedule added to the challenge of preparing good experiments. Although the *Apollo Experiments Guide* did not include science payload weight allocations, we continued to plan based on 250 pounds. We divided this weight into three parts: 100 to 150 pounds reserved for a surface geophysical station, 100 pounds for the geology equipment, including cameras and sample containers, and a small allocation for orbital science, essentially whatever might be left over. When potential experimenters inquired about payload availability, we offered these numbers for planning their submissions.

At the end of September 1965, in response to a request by Bob Seamans and as an elaboration on the plan we circulated in June, Mueller and Newell forwarded the first "Lunar Exploration Plan."[11] The forwarding memo stated that the attached plan had been coordinated between OMSF and OSSA. This was indeed true, for along with others I had worked on the attachment wearing both my OMSF and OSSA hats. Events were moving rapidly, however, and during the three days between completing the plan and sending it on to Seamans, two major management decisions had been made: Surveyor missions after *Surveyor 6* and Lunar Orbiter flights after *Orbiter 5* would be canceled. We went back to modify the plan reflecting these changes, and at the end of October we issued a revised plan noting that there might be follow-ons to the Surveyor and Lunar Orbiter programs after 1970, though no funding was identified. Seven Apollo missions, including test flights and the first landing attempts, were shown on the schedule through 1969, and by the end of 1971 these would be followed by three Apollo Applications Program (AAP) surface missions and three orbital missions. Additional AAP surface and orbital missions were dashed in on the schedule chart through 1973, and after that date a new category, Extended Manned Missions, would begin, continuing beyond 1975.

From our perspective this plan contained all the right words, words we had labored to have our senior management embrace publicly for the past two years. Now we had it in writing. To give just a brief sample, the plan stated: "The primary objective . . . is to define the nature, origin, and history of the moon as the initial step in the comparative study of the planets. . . . A secondary objective, following naturally from the first, is to evaluate the potential uses of the moon." Apollo and post-Apollo lunar exploration would accomplish all we wanted if the words were followed up with action. But only NASA management had bought into the plan; allies in the administration and Congress were still lacking. The plan would be updated from time to time, not always by formal documents but by working papers written to reflect the latest guidance and the realities of NASA funding projections.

To improve our relationship with the MSC Flight Operations Directorate (FOD) and benefit from its "real mission" experience, we invited some of the flight controllers to come to Flagstaff and witness a training exercise we would be conducting for a post-Apollo mission simulation. Our demonstration of Command Data Reception and Analysis, a smoothly functioning embryonic science support room, once denigrated by MSC, convinced FOD that an experiments room would be a valuable asset.

After much give and take on how experimenters and the science community would interact with mission controllers and the astronauts in real time during an Apollo mission, MSC agreed in 1967 to build an experiments room in the mission control building. Christopher Kraft and his flight controllers in FOD deserve the credit for recognizing the wisdom of having such a facility, but the intervention of Jack Schmitt, Donald Lind, and other astronauts who had worked with the training and simulation teams assembled by USGS was critical to getting this agreement. They had firsthand knowledge of how valuable it would be for the crews on the lunar surface to have experienced scientists backing them up.

The arrangement was formalized in April 1967, when FOD issued its "Flight Control Handbook for Experimenters."[12] It called for an experiments room, later named Science Support Room (SSR), to be located in building 30 near the Mission Operations Control Room (MOCR). The MOCR was the large room, filled with banks of monitors manned by engineers in short-sleeved white shirts and ties, seen by everyone who watched the Apollo space missions on television. During initial discussions it was proposed that the experiments room be located with other support teams in building 226, a few blocks away, and for *Apollo 8* that was its location. However, we were able to convince Chris Kraft that for the landing missions it had to be nearer the action, like other critical Staff Support Rooms (SSR again), so that the displays and other information we planned to coordinate would be accessible to those who might have to make quick decisions. This would be especially important for the later missions, when we expected that lunar surface operations would be much more complex and timelines would be jammed with tasks. Being in the same building as the MOCR also let us use the pneumatic tube message system that connected all the SSRs in the Mission Operations building and was used extensively to pass information around. This sounds primitive today, when it is so easy to communicate between computer terminals, but in 1967 it was state of the art and local area networks were still a technology of the future. The staffing and layout for the experiments room were still under study at the time the handbook was issued, but eventually we were assigned room 314, which contained TV monitors, tables, phones, other equipment, and eventually closed-circuit television that allowed quick exchange of vital information. Perhaps as a small bone to keep the headquarters types off their backs, a console was designated for a headquarters representative, and that is where we usually were stationed when the missions began rotating shifts with Ed Davin, John "Jack" Hanley, Donald Senich, and me.

In the coming years, as we continued to refine our activities in the SSR, it became clear that we needed more space to accommodate all the people and equipment we required to follow the action. Another small SSR was added in the building; Raymond Batson from USGS recalls that during *Apollo 11* this auxiliary SSR got so crowded you could hardly move around. In addition to Ray's crew, who were monitoring the television pictures coming back from the Moon and the air-to-ground conversations with the astronauts, Bendix engineers were at their consoles keeping track of the data transmitted from the deployed experiments. Court reporters were also taking down the voice communications so this historic record wouldn't be lost if the tape recorders malfunctioned, as they frequently did in NASA's early days.[13] After *Apollo 11* the auxiliary SSR was moved to a larger room where a plotter allowed Ray's crew to create a real-time map of each landing site showing where the astronauts were and had been. They would supplement the map with Polaroid panoramas captured from the TV pictures sent back to Earth. Based on all this information, the staff and PIs in the SSRs would formulate questions and send them to the capsule communicator (CapCom), who would then decide whether to pass them on to the astronauts.[14] Later in the program, for the final landings, three SSRs were staffed, two for surface science and one for orbital science.

As soon as a *Saturn V* cleared the launch tower, control of the mission transferred from KSC to MSC. MSFC also continued to play an important role throughout the mission and kept a crew at MSC, since they were the experts to be consulted if there were problems with any of the Saturn rocket stages. Backing up the SSRs would be support rooms in building 45 for all of Apollo's major systems. They were manned by contractor and NASA staff who had access to detailed knowledge of what made the systems and experiments tick.

This behind-the-scenes support, which most people who followed the missions were unaware of, figured prominently in saving the *Apollo 13* astronauts and was portrayed rather accurately in the movie. Every detail for every system and subsystem could be found and displayed in these rooms, almost instantly, and they were manned around the clock while missions were under way. They were connected by phone to the MOCR and in most cases were directly linked to the contractor's plant or manufacturing facility so that additional brainpower could be brought to bear in an emergency.

As important as it was for the experiments to have assigned SSRs, the handbook also formalized the procedures for simulations with the flight controllers. This was another major step forward and for the first time placed experiment simulation in the mainstream with all the other simulations carried out for the missions. Simulations would cover normal and abnormal situations that might require consultation with the SSR, and the flight controllers were given particularly wicked problems as they gained experience. The schedule called for the experiment simulations to start four weeks before launch, so beginning in June 1969 we had to man the SSR with the staff that would be present during the actual missions.

A memo to my staff in September 1970 lists a schedule for *Apollo 14* surface experiment simulations, giving an idea of what these simulations entailed.[15] By this time simulations were conducted from the Mission Control Center, Houston (same place as MOCR, different name). The memo called for two simulations of the planned first EVA and three simulations of the second, spread over two months rather than the one month originally planned. It was getting hard to assemble the large cast of characters that was required and, more important, to fit the simulation into the astronauts' tight schedules. The simulations would include the prime crew, using either sites at KSC or one designated by Flagstaff. There were also two "canned" simulations at Houston when the astronauts were not part of the exercise and the flight controllers and our SSR staff were tested with contrived problems. Later missions, because of their complexity, added additional simulations. Each simulation would last four hours or more and would be followed by a candid critique, usually leading to new guidelines on how to respond to emergencies during the real mission.

As the PIs and their supporters began to spend more and more time at MSC, the members of the Field Geology Team availed themselves of a rather unusual perk. Jack Schmitt had long since completed his flight training and was now in Houston full time. He had a modest bachelor apartment just a few blocks from the center. His old Flagstaff buddies saw nothing wrong in staying there when they were in town, and if you visited Jack late at night you usually found at least one of them in a sleeping bag on the floor. I don't know how many keys were in circulation, but Jack's hospitality helped the visiting team members stretch their meager government per diem to include extra dinners at the San Jacinto Inn, the Rendezvous, or some other favorite restaurant. Jack was also using the LM and CSM simulators at MSC and KSC when they were not scheduled for designated crew simulations, to become familiar with these complicated spacecraft. When Jack was selected in the first scientist-astronaut class in 1965, some of us who knew him at Flagstaff recommended that he make it clear to Deke Slayton and Al Shepard how seriously he wanted to be looked on as one of the "regular guys," removing any stigma from his hyphenated title. Whether or not this urging had any influence, Jack spent long hours in the simulators and added to his flight log by flying the astronauts' T-38s around the country, frequently coming to Washington to attend meetings and briefings at headquarters. Did Jack's diligence have any direct effect on Slayton and Shepard? I have to believe it did, and as we know, he was selected for the crew of the final Apollo landing mission.

Mission Control interactions with the experiments to be conducted on the journey to the Moon or on the way back home, as well as those conducted in lunar orbit, were not completely defined in 1967, but the groundwork had been established. Each experiment was assigned an FOD experiments activity officer who would represent the experiment through all phases from planning to flight operations. This person would work with the PI(s) to ensure that the experiment was properly integrated and operated. If a mission contingency should arise requiring some modification to normal operations, the EAO was charged with coordinating with the PI and then representing his interests in maintaining the experiment's integrity during the brainstorming to solve the problem. Although it sounds bureaucratic, acknowledgment that such interaction might be necessary was another encouraging sign that science objectives had moved up in the MSC engineering culture. With so much going on during a mission, great discipline was required for all mission operations, and precise procedures were followed for all the flight systems—not just the experiments—during the actual missions. But by the time the Apollo flights began, PI relations with the flight controllers had improved significantly, and minor adjustments could be made in a much less formal atmosphere. Most of the FOD staff became strong champions for science, and when obstacles arose they did all they could to overcome them.

Another advance for science was the promotion of scientist-astronauts to be mission scientists and CapComs during the lunar landing missions. CapComs were the only ones allowed to speak directly to the astronauts during missions, and they had to be astronauts themselves, a rule still followed for all manned missions. This is not to say that the other astronauts serving as CapComs did not do an acceptable job in directing the crews or relaying information and suggestions to them. But this change went a long way toward reassuring us, especially the field geology PI, that the best advice would be quickly available if the astronauts met with some unexpected discovery or predicament on the lunar surface. We had always hoped that the PIs, and other Earth-bound scientists, would be able to communicate directly with the astronauts, but this never happened except for one instance described in chapter 12.

In mid-September 1967 I attended a dry run at MSC of a session on Apollo mission planning that would be presented later to MSC senior management.[16] Owen Maynard of the Apollo Spacecraft Project Office (ASPO) chaired the meeting. Maynard had been involved with Apollo from its earliest days, having served in 1960 on the Langley Space Task Group that drew up the first specifications for the launch vehicle and Apollo spacecraft. With Joe Shea, he had enumerated the steps that had to be achieved as the program progressed toward a lunar landing. At this meeting we were briefed for the first time on the development schedule that MSC expected to follow leading up to the first landing, which was now designated the G mission.[17] Joseph Loftus discussed the three types of missions that were possible when we reached the final level: (1) touch and go—this mission might stay on the lunar surface for as little as two hours with no EVA permitted, have an umbilical EVA of half an hour, or have an EVA of an hour and a half with the astronauts using the portable life-support system (PLSS) within a limited radius of the LM; (2) limited stay—structured around twenty-two and a half hours on the lunar surface, one EVA, and no deployment of the Apollo Lunar Surface Experiments Package (ALSEP), an automated geophysical laboratory or ground station; and (3) maximum stay—with four EVAs, each lasting up to three hours.

During discussion of these three options, ASPO made it known that it favored the limited stay mission for the first landing. Thomas Stafford, representing the astronaut office, pointed out that on the Mercury and Gemini flights it was only after the fourth flight that the spacecraft became really operational, and he expected the same for the LM. He mentioned that LM propellant leaks might restrict the surface staytime and said he thought this situation would improve as LM production continued. He also was concerned that with all the other high priority training they would need, the crew for the G mission would have a hard time completing the required training to carry out a multi-EVA mission. For these reasons he also supported the limited stay as the best that could be accomplished on the first landing. A few days later, at the MSC directors' briefing, the limited stay mission was endorsed with one modification; ALSEP deployment would not be deleted. Thus, some two years from the date the first landing would be scheduled, we saw that planning for man's first lunar landing would continue to follow a conservative mission profile. A small victory at the time, ALSEP would still be a part of the science payload.

Soon after this decision was announced, the MSC Crew Systems Division began regular monthly meetings to review and highlight any new problems that could affect the astronauts' EVAs. This new group was named the Lunar Surface Operations Planning Committee and was chaired by Raymond Zedekar. The meetings were well attended by the various MSC offices that had a finger in any of the EVAs. We had established a good working relationship with Ray, so our office was invited to attend as well as staff from Bellcomm and USGS.[18] These meetings covered a wide range of topics, including the latest results of space suit simulations and their implications for the astronauts' ability to perform certain types of surface tasks, and we reviewed all other EVA concerns such as PLSS power budgets, tool design, and sampling procedures. These meetings continued through 1968 and were later replaced by another planning process.

As 1967 was winding down and we were assimilating the advice we received at Santa Cruz, the last major organizational change involving Apollo science was made at NASA headquarters. Still wearing my two hats but officially assigned to the Advanced Manned Missions Program Manned Lunar Missions office, in early December I was moved to a staff position in anticipation of a new assignment.[19] By the end of the month, Mueller established the Apollo Lunar Exploration Office, reporting to Sam Phillips, and put Lee Scherer in charge.[20] Lee had just finished tying up the loose ends from the Lunar Orbiter program, and this appointment gave him a chance to expand his management role. His new office combined the responsibilities of Foster's office and some of the post-Apollo lunar exploration duties of Advanced Manned Missions. He inherited most of Foster's staff as well as other headquarters staff who had become involved in lunar science, including William "O.B." O'Bryant and Richard Green. They had been managing the development of the Apollo geophysical station (ALSEP) in the Office of Space Science and Applications. As part of the agreement to establish this new office, OSSA continued to fund the lunar programs it had started through the end of FY 1969. O'Bryant was named assistant director for flight systems and continued to be in charge of ALSEP. Noel Hinners and his growing Bellcomm group also switched hats and supported our new office. Will Foster was given a staff position within OSSA to oversee Apollo experiment selection.

Scherer's appointment was a management masterstroke by Mueller. He was well liked and trusted by John Naugle (who had replaced Homer Newell just three months earlier) and by the science side of NASA, having managed the highly successful Lunar Orbiter program. The close connection of Lunar Orbiter to Apollo made him well known to OMSF management. After our initial meeting in 1963, I got to know him well from working with his NASA and contractor team during Lunar Orbiter site selection meetings. Perhaps it was his navy connection and my familiarity with the navy way of doing business, but with his appointment I expected to see more progress in all aspects of Apollo science. Lee would have much greater influence on the decision makers than Will Foster did. Being on Phillips's staff put him directly in the chain of command—no more half OSSA and half OMSF, with both offices never sure whose side you were on. We were all now, clearly, part of the Apollo team. Most of the senior NASA managers on Apollo were either active-duty or retired military officers, so Lee fit right in. With my new office colleagues I had a change of address and moved into the Apollo offices at the just completed L'Enfant Plaza complex, where we remained until the last mission came home. I was given a new title in Scherer's office—program manager, plans and objectives. My new responsibilities involved me in all aspects of Apollo science; most important was the planning for what would come after the first few flights.

The Apollo program was overseen by several special committees; perhaps the most prestigious was OMSF's Scientific and Technology Advisory Committee (STAC). Its membership comprised distinguished scientists and engineers. Chaired by Charles H. Townes from the University of California, Berkeley, it was increasingly important as Apollo neared its first launch. It met quarterly with Mueller and other senior NASA management to review all aspects of the program. At the beginning of April 1968, Townes wrote to Jim Webb expressing the committee's satisfaction with the program's status and also its concerns.[21] He stated that after spending seven days reviewing various steps in the mission, the committee believed that "NASA personnel involved in this effort are mastering well a very demanding and difficult, as well as an exciting, assignment." He wrote, however, that "it did not appear that efforts toward working out operational procedures for activities on the moon and coordinating the astronauts' abilities and restrictions with optimum scientific experimentation had yet made comparable progress." And in referring to the NASA budget reductions, Townes closed with, "We believe it would be poor economy indeed for the nation to jeopardize the chances of a ringing success for the entire effort by undue paring down of support during the last stages which are ahead." STAC's concerns echoed those being expressed by our new office, and I believe they went a long way toward elevating Lee Scherer's influence with Apollo management in the months leading up to the first landing.

At the beginning of 1968 our office prepared to update the 1965 "Lunar Exploration Plan." A Bellcomm technical memorandum written in January also addressed long-range lunar exploration planning.[22] It was distributed widely inside and outside NASA with the purpose of justifying a continuing program of exploration after the Apollo landings and rebutting the recently announced reduction in FY 1969 funding that would discontinue missions after *Apollo 20.*

The memo outlined a program based on the Bellcomm authors' judgment of the scientific results that would be achieved by exploring specific sites using lunar orbital surveys and on our AAP concept of using a rendezvous between an extended lunar module and an unmanned LM payload module to permit longer staytimes and greater payloads. Except for listing the landing sites they thought were most important and giving their rationale for choosing them, their memo did not propose any major changes in previously circulated internal documents describing AAP plans. The memo placed Bellcomm management squarely on our side in support of dual-launch missions. Until this time it had only gingerly endorsed the approach we had been advocating for several years in the Advanced Manned Missions office.

At the time the Bellcomm memo was circulating, a senior NASA management team called the Planning Steering Group was put in place to furnish an overall NASA stamp of approval for the agency's long-range space exploration plans. In April 1968 Scherer established a Lunar Exploration Working Group to reexamine the situation and recommend a long-range exploration program to the PSG. He hoped to influence the NASA FY 1970 budget proposal and perhaps change the administration's mind about what needed to be done after the initial landings. The Lunar Exploration Working Group included members from MSC, MSFC, Langley Research Center, JPL, and Goddard Space Flight Center in addition to headquarters. John Hodge of MSC was appointed director of the effort. We met frequently during the spring and summer of 1968. George Esenwein, Martin Molloy (detailed from JPL), and I took the lead for Scherer's office. We had many differences of opinion with the MSC representatives on the working group concerning what should constitute a long-range lunar exploration plan, especially in regard to using dual launches to extend staytime and permit greater science payloads.[23] But eventually, reinforced by the recommendations of the Santa Cruz summer conference and by the Bellcomm report, we prevailed and shaped a program similar to the one we had proposed earlier for AAP.

In October 1968 we distributed a Program Memorandum for Lunar Exploration.[24] With funding constraints uppermost in our minds, we tried to throw the ball back to the Bureau of the Budget by quoting from and answering an earlier BOB inquiry: "What program should be undertaken for lunar exploration after the first manned lunar landing?" Our memorandum outlined such a program, and to give it additional clout, we also quoted from a 1963 President's Science Advisory Committee (PSAC) report and the 1965 study by the National Academy of Sciences. Both had made strong statements that continued lunar exploration was essential to unraveling important scientific questions. This memorandum, like the 1965 plan, proposed an exploration program that would extend beyond 1975. It included manned and automated missions, dual launches, and even new hardware systems. The guidance we had received from BOB for our FY 1970 submittal was that NASA should pause after the first few landings and wait some unspecified time before continuing lunar exploration. (Typically BOB issued guidance each spring for drawing up each agency's budget for the next year. This guidance included the language and dollar targets it expected the agencies to adhere to when they submitted their budget requests to the administration later in the year.) Between 1963, when we quoted PSAC's opinions on the importance of exploring the Moon, and 1967 a major shift had occurred. PSAC's new view was that "repetition of Apollo flights for more than two or three missions will be unjustifiable in terms of scientific return without the modification of the system to provide for additional mobility. . . . and the capacity to remain on the surface for a longer period of time." We could not have agreed more. Unfortunately, without a budget increase, what PSAC was suggesting couldn't be done.

The final pages of our memorandum addressed these issues. We rejected the option of pausing, for several reasons, and proposed that either we continue without modifying the Apollo hardware, in order to maintain momentum, or start to modify the basic systems to improve the astronauts' mobility and extend staytime. If either of these last two options was accepted, we would need additional funding in FY 1970. BOB rejected our request for more funds but eventually permitted NASA management to juggle the approved budget and make the changes that resulted in the J missions to be discussed in following chapters.

At the end of the Santa Cruz conference, in the summer of 1967, Bill Hess established an interdisciplinary Group for Lunar Exploration Planning. Its objective was to integrate the science planning for each mission and offer an overall strategy to ensure that the missions complemented each other for the maximum scientific return. With the AAP missions at least on hold, GLEP focused on coordinating the planning for the Apollo missions. Planning centered mainly on selecting landing sites. Each site's unique characteristics would dictate the experiments to be carried out and how the geological surveys would be conducted.

To do the staff work in support of GLEP, a small group of scientists and engineers that we dubbed the "rump GLEP" met to put all the pieces together for presentation to GLEP. The rump GLEP initially included (besides me) Hal Masursky and Don Wilhelms from USGS; John Dietrich and John "Jack" Sevier from MSC, joined at times by Jack Schmitt; several scientists from outside NASA, including Paul Gast and Eugene Simmons; and two Bellcomm staffers, Farouk El Baz and Noel Hinners, the latter chairing the meetings. For the next two years we met regularly to plan each of the upcoming flights, updating our recommendations as more and more information became available. We were not the only ones trying to identify landing sites; many others at MSC and Bellcomm besides those mentioned above were also putting in suggestions. But because of our diverse backgrounds and intimate knowledge of mission constraints, we felt we were the only team working on candidate sites that had the big science and operational picture in mind.

The site selection process involved making recommendations to GLEP accompanied by supporting arguments. Based on this work, lists periodically went to GLEP adding or subtracting sites as advocates made the case for one site or another. GLEP, in turn, would make recommendations to ASSB, the final arbiter in site selection. Work on selecting landing sites became more intensive as the launch dates drew nearer. The few sites finally chosen would represent the coming together of many interests, both scientific and engineering. If someone held a strong position or theory on some aspect of lunar science, you would hear arguments for sites that held the most promise of vindicating that position. Site politics could rear its head at times; but fortunately consensus prevailed, though for several landings we chased the ephemeral "recent volcanics" advocated by a small USGS clique and others. Many people spent long hours reviewing the Lunar Orbiter photographs and other information to arrive at the recommended sites. As Noel Hinners's staff gained strength with the addition of James Head and others, they worked closely with USGS in Menlo Park and Flagstaff and took the lead in providing site rationale for GLEP. The importance of selecting the right sites could not be overestimated: they would shape and control our understanding of the Moon for many years to come.

For the first landings, Lunar Orbiter photography, supplemented by USGS

1:1,000,000 scale lunar quadrangle geologic maps made from telescopic studies, were the key sources we used to develop a list of recommended landing sites. Lunar Orbiter coverage was designed to supply the following products for the initial landing sites: a series of photographs with three-foot ground resolution; detection of obstructions eighteen inches high; stereo coverage for detection of slopes of seven degrees or greater; approach path coverage of the last twenty miles of the LM approach to the landing site; and oblique views to approximate what the LM pilot would see as he approached the landing site. We selected thirty-two sites in the "Apollo zone" that met these specifications, and they were designated set A. We then turned these sites over to the Mapping Sciences Branch at MSC for final "landability" analysis.[25]

From set A, eight sites (set B) were selected that incorporated all the landing site considerations, including proper lighting and separation to allow three launch attempts, two days apart, in case of launch-pad holds. This last constraint was imposed to avoid costly detanking (removing the propellants), and rechecks of all the Apollo systems if the launch to a selected site was missed for any of several possible reasons. If no secondary or tertiary landing sites were available, a launch abort would require a month's delay to arrange lighting at the initial site for avoiding obstacles. For the first landing attempt, set B was further refined to a five-site set C that included Tranquility Base, *Apollo 11*'s final destination. *Apollo 12*'s site, near *Surveyor 3*, was included in set B.

In March 1968 President Johnson announced the formation of the Lunar Science Institute (LSI). The National Academy of Sciences had pushed such an institute to offset the continuing perception by many in the scientific community that NASA was not paying enough attention to science on Apollo. The site selected was a renovated mansion belonging to Rice University, just outside the MSC fence. William W. Rubey, one of the renowned scientists who had volunteered time to work with the astronauts during their early training, was appointed the first director. Still on the faculty at the University of California at the time of his appointment, he was a popular choice and gave the institute instant credibility.

At headquarters we supported the need for the institute but were not keen on the location. We felt that MSC's proximity and reputation might discourage scientists from taking advantage of the institute's mission to provide a base from which to work on the material and data the Apollo flights would return. Other purposes, such as attracting graduate students and scientists on sabbaticals and hosting conferences and seminars, might also suffer because of the climate of distrust that existed. These fears went away in the ensuing years as LSI (later named the Lunar and Planetary Institute) ably performed its functions and remained independent of MSC.

Although LSI was chartered by the National Academy of Sciences and its board of governors was appointed by the Academy, most of the funding came from the Apollo program.[26] Eventually LSI outgrew its initial home and moved to more spacious quarters at Clear Lake, where it continues to be a focal point for the study of Apollo material as well as information returned from later lunar and planetary programs.

The crater Aristarchus is the brightest feature on the Moon's Earth-facing surface and the site of many telescopic observations of color changes and other phenomena. (NASA AS-515-2609)

Northrup post-Apollo lunar drill engineering model, on test stand at Marshall Space Flight Center, demonstrates deployment of coiled drill string and core stem. (NASA MT 6-9401) *Inset,* artist drawing of drill attached to Lunar Module. (MS-G-114-1-65; drawing courtesy of John Bensko)

Philip Culbertson, NASA headquarters manager, at the controls of a Bendix model local scientific survey module during a demonstration in Flagstaff, Arizona. (Photo by Philip Culbertson)

Gordon Swann simulates geology field operations at Flagstaff, Arizona, while wearing a prototype Apollo space suit. (USGS photo)

The post-Apollo two-man test vehicle (MOLAB) arrives at Flagstaff, Arizona. (USGS photo)

Space-suited test of prototypical geological equipment at the crater field simulation site near Flagstaff, Arizona. (USGS photo)

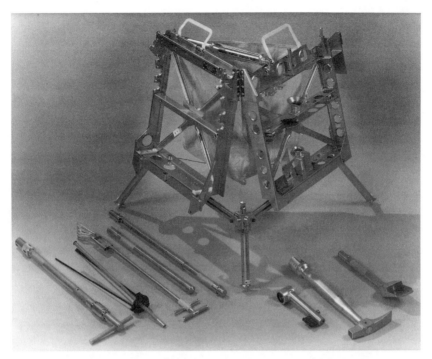

The tool carrier used on *Apollo 12* and *Apollo 14* includes a partial tool inventory: scoop, hammer, magnifying lens/brush, two core tubes, tongs, gnomon, and extension handle. (NASA S69-31867)

Astronauts carried two Apollo lunar surface return containers, or "rock boxes," on all missions, sealing them on the Moon to protect and preserve the samples collected. (NASA S72-37196)

NASA engineer John B. Slight climbs in a simulated lunar crater at the Manned Spacecraft Center, Houston, while the six degrees of freedom test rig provides a "feel" for the Moon's one-sixth gravity. (*Top*, NASA 65-H-218; *bottom*, NASA 65-H-633)

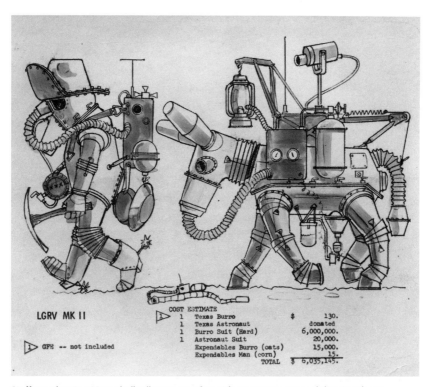

Staff member Bart Campbell's illustration of an aide to astronauts' mobility. (Author's collection)

Astronaut Walter R. Cunningham tests a prototype Apollo space suit at Bend, Oregon, in August 1964. (NASA S64-31319)

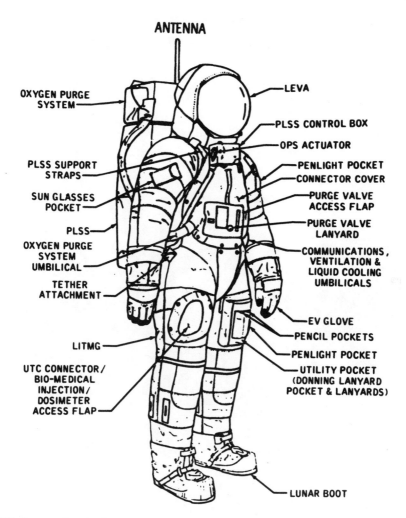

ANTENNA

OXYGEN PURGE
SYSTEM

LEVA

PLSS CONTROL BOX

OPS ACTUATOR

PLSS SUPPORT
STRAPS

PENLIGHT POCKET

CONNECTOR COVER

SUN GLASSES
POCKET

PURGE VALVE
ACCESS FLAP

PLSS

PURGE VALVE
LANYARD

OXYGEN PURGE
SYSTEM
UMBILICAL

COMMUNICATIONS,
VENTILATION &
LIQUID COOLING
UMBILICALS

TETHER
ATTACHMENT

EV GLOVE

PENCIL POCKETS

LITMG

PENLIGHT POCKET

UTC CONNECTOR/
BIO-MEDICAL
INJECTION/
DOSIMETER
ACCESS FLAP

UTILITY POCKET
(DONNING LANYARD
POCKET & LANYARDS)

LUNAR BOOT

This diagram of an Apollo space suit shows the various components that made up the final configuration. (From NASA EMU Data Book)

Textron Bell Aerosystems engineer preparing to test a Lunar Flying Vehicle on 1/6 g test rig, Buffalo, N.Y., June 1966. (Photo courtesy Niagara Aerospace Museum)

Top, astronaut training trip, Meteor Crater, Ariz., 19 May 1967. (Photo courtesy George Ulrich, USGS.) *Bottom,* aerial view of Meteor Crater. (Author's files)

Astronauts Alan B. Shepard Jr., *left,* and Edgar D. Mitchell practice geology tasks for their *Apollo 14* mission using a small two-wheeled cart (modularized equipment transporter, MET) to carry cameras, geology tools, and sample bags. (NASA S70-54167)

Astronaut David R. Scott, with camera, and scientist-astronaut Joseph P. Allen, *Apollo 15* CapCom, during a Flagstaff simulation. Scott practiced his lunar standup EVA from the LM mockup in the background. (NASA S70-53280)

Astronauts David R. Scott and James B. Irwin practice with the lunar vehicle simulator at the Flagstaff crater field. (NASA S70-53299)

Astronaut David R. Scott operates the *Apollo 15* lunar drill during a Kennedy Space Center simulation. (NASA S70-29673)

Astronauts John W. Young, *right,* and Charles M. Duke, *center,* stand behind the Explorer test vehicle used for an *Apollo 16* crew simulation. (NASA S72-31183)

APOLLO 17
SIM BAY

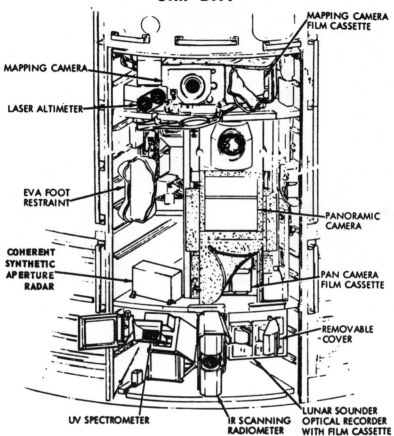

MAPPING CAMERA FILM CASSETTE

MAPPING CAMERA

LASER ALTIMETER

EVA FOOT RESTRAINT

PANORAMIC CAMERA

COHERENT SYNTHETIC APERTURE RADAR

PAN CAMERA FILM CASSETTE

REMOVABLE COVER

UV SPECTROMETER

IR SCANNING RADIOMETER

LUNAR SOUNDER OPTICAL RECORDER WITH FILM CASSETTE

Line drawing of a scientific instrument module (SIM) bay showing equipment installation in a bay of the *Apollo 17* service module. (from NASA Apollo 17 Press Kit)

During this simulation, conducted in the water immersion, neutral buoyancy facility at the Manned Spacecraft Center, Houston, an astronaut removes the film canister from the SIM bay. (NASA S71-58149)

6

Developing the Geological Equipment, Related Experiments, and Sampling Protocols

Methods of conducting geological field studies have changed little in the past two hundred years. The geologist visits the locale to be studied, samples rocks, measures structural features like hills, valleys, cliffs, and other surface topography, traces formation boundaries (if possible), determines the relative ages of these various features, usually by several techniques, then interprets this information and finally makes a map. Aerial and satellite photos, as well as new surveying instruments and global positioning systems, now simplify and speed up the fieldwork, but all these steps are still necessary to produce a final map. In many cases geophysical data can help in making subsurface interpretations, but the overall job remains the same: sample, measure, interpret. Depending on the geological complexity of the site and the geologist's skills, this can be a time-consuming endeavor. Some sites have been studied for years by the same or different geologists, slowly yielding an interpretation that most workers will agree with.

Lunar geological fieldwork would present the same challenges that faced a terrestrial geologist plus many more. For example, at the beginning of Project Apollo it was not clear how easily astronauts could sample and measure lunar features; above all, in spite of the many hours spent in geology training, it was questionable how skilled they would be at deciding how and where to sample and take measurements. Each Apollo landing site would represent a one-shot opportunity to collect as much information as possible—there would probably be no return to resample or remeasure—so it had to be done right. This demand haunted the new breed of "lunar geologists": they had to complete the job the first time. That very little hard data would be in hand until the Apollo landings took place (Ranger, Surveyor, Lunar Orbiter, and ground-based observations notwithstanding) added enormous complications for those of us attempting to prepare the equipment that would be taken on each mission and to plan the exploration strategy.

In February 1964 Will Foster sent a set of recommended Apollo investigations and investigators to the Space Science Steering Committee (SSSC),[1] the group Homer Newell had charged with advising him about what science to conduct on all space programs. In his memo Foster listed five areas of Apollo investigations—geology, geochemistry, geophysics, biology, and lunar atmosphere—and named scientists who should be on the investigating teams. As expected, the recommended geology fieldwork team was headed by Gene Shoemaker. It included Hoover Mackin from the University of Texas, Aaron Waters from the University of California, Santa Barbara, and Edward Goddard from the University of Michigan. The geochemistry planning panel included James Arnold from the University of California, San Diego, Paul Gast, then at the University of Minnesota, Brian Mason from the American Museum of National History, and several other noted geochemists. Related to the geochemistry panel was the petrography and mineralogy team composed of Harry Hess of Princeton, Clifford Frondel of Harvard, Bill Pecora and Ed Chao of the United States Geological Survey, and Edward Cameron of the University of Wisconsin.

Shoemaker's Field Geology Team was responsible for planning the lunar fieldwork, determining the requirements for maps and tools, monitoring the astronauts' training and their activities once they reached the Moon, and preparing the necessary reports. Working with the geochemistry planning panel and the petrography and mineralogy team, the Field Geology Team would plan sample collecting procedures and design sampling equipment that would satisfy the needs of future sample-analysis PIs. For samples that would be returned to Earth, the geochemistry planning panel and the petrography and mineralogy team would recommend the protocols for sample preparation. Finally, the geochemistry planning panel was asked to recommend to Foster's office particular investigations and investigators for studying the samples. These teams and panels were subsequently approved by the SSSC and began their work.

Before Shoemaker's appointment, two conflicting concepts for field geology instrumentation were under development, one designed by the staff at the Manned Spacecraft Center and the other by USGS in Flagstaff. MSC, led by Uel Clanton, had devised an engineering model of an all-in-one geological tool that the astronauts could use for sampling, drilling, and several other functions, in an attempt to simplify the many tasks they would have to accomplish and at the same time save weight and time by reducing the number of tools needed.

USGS had similar concerns but thought the biggest problem would be locating and documenting the sites visited, and in particular sampled, so that accurate traverse maps and profiles could be reconstructed back on Earth. The Flagstaff team had devised a surveying staff that would reflect a laser beam from a ranging device and automatically record the coordinates of a position on the lunar surface. This approach was based on the simulations and exercises we had been conducting for the post-Apollo missions, which suggested that without some type of surveying instrument it would be almost impossible for an astronaut to accurately locate his position on the Moon and associate a sample or observation with a specific point. Lunar geologic maps made without such positioning would be seriously degraded in value, since to establish map locations we would have to depend on some type of dead reckoning or coarse Earth-tracking and reconstruction of the traverse based on voice communication.[2]

Our experience during the Martin Marietta contract, and the growing concern about measuring distances on the lunar surface, led the Branch of Astrogeology to further explore including a periscope in the lunar module (LM), as we had proposed earlier, rather than the sextant that was being planned for navigation. In February 1965 Gordon Swann and Dave Dodgen visited two navy periscope suppliers, Kollmorgan and Kollsman Instruments, to discuss their ideas. Besides the concerns arising from the Martin contract, they wanted to be able to track an astronaut if only one was allowed to leave the LM. Though both companies thought the Apollo navigation requirements and the surveying ability needed on the Moon's surface could be incorporated in one instrument,[3] no official action was taken. A jury-rigged optical ranging periscope built by David Dodgen and Walt Fahey was used during some field simulations to assess the value of such an instrument.

These three pieces of equipment had their advocates and their detractors. At the end of 1965 the MSC engineering model was tested by a joint review team composed of members of Foster's office and several MSC offices, including representatives from the astronaut office, and we agreed to stop work on this tool. Because of its several functions, it was large and cumbersome, with so many batteries, handles, switches, and other components that it looked like a Rube Goldberg contraption. The USGS surveying staff survived our initial evaluations. In spite of the advertised versatility of these tools, the astronauts would still need additional equipment for tasks that the all-in-one designs could not perform. Converting the LM sextant to a periscope was also finally abandoned because of the added cost and schedule delay entailed by modifying the LM navigation system. For the last three missions, a navigation system on the astronauts' lunar rover met most of the tracking and mapping requirements.

As we began to design and build prototype tools, another complication arose: certain materials and designs might interact dangerously with the spacecraft's atmosphere, communications, or even the astronauts' space suits. These restrictions, some certainly necessary, would be a bone of contention throughout the equipment development phase, adding trouble and expense to what could have been, in some cases, rather straightforward procurements.

Without question, the most important task the astronauts would perform on the lunar surface would be sample collection. There was much debate on how best to do this. How much sample? What types of samples? How should they be packaged for the trip home? How badly would the lunar surface, and in turn the samples, be contaminated by the effluents from the LM descent engine plume? These questions and many more faced us as we began to realize that a lunar landing was not far off. The danger of contaminating the Earth was being addressed, but designing the sample containers to minimize this concern still lay in the future. Answers to all these questions would affect the design not only of the sample containers but also of the collecting tools.

To start answering the sampling questions, the Office of Space Science and Applications asked USGS to detail to NASA a person with experience in sample collection and analysis. Ed Chao was the first to arrive, soon followed by Verl Richard Wilmarth, a senior USGS manager. Dick arrived at NASA in early 1964, and I first met him soon afterward in his new office in federal office building 6. NASA shared FOB-6 at that time with other government agencies, and though it was older than FOB-10, where my office was, the building was more luxurious; wider corridors, bigger elevators, a fancier cafeteria, and the other trappings of power so important in Washington. The NASA administrator and senior staff had offices in this building as well as OSSA, the General Council, Legislative Affairs, Public Affairs, and several other NASA departments. The top floors had been taken over by NASA, and some offices afforded a wonderful view of the city. The administrator's office faced west toward the

White House, and Legislative Affairs looked east toward Capitol Hill—perhaps by some logic, though probably just by chance.

Although he was an experienced manager, Wilmarth had never had an assignment quite like this: soliciting the scientists of the world to bid for a piece of the returned lunar samples and perhaps a chance to win a Nobel Prize—a once in a lifetime opportunity. I told Dick about my experience in developing this type of solicitation, officially called an announcement of flight opportunities (AFO), as well as my background in writing government requests for proposals (RFPs) that had been released from NASA headquarters. Lacking this experience, especially with the quirks of NASA procurements, he asked me to assist him in his new job.

For the next several months Dick wrestled with his task, and I spent a significant part of my time helping him. Many meetings and consultations with interested parties were needed to be sure we were not overlooking some large or small detail. The AFO had to ask for information covering several areas, in a form that would let a blue-ribbon panel, still to be identified, select the most qualified proposals. What was the objective of the analysis? How much sample was needed? Would the analysis involve destructive or nondestructive testing? What were the packaging requirements? What type of equipment would be used? Would there be collaborators in addition to the principal investigator (PI), and who would they be? How much funding would be needed? How long would it take to do the analysis? Finally, after several months of labor, a draft of the AFO was ready to be circulated to senior management, and after review by both OSSA and the Office of Manned Space Flight, a final version was released at the end of 1964. The AFO asked that proposals be delivered to NASA by June 1965.

Before the sample proposals were received, Shoemaker's Field Geology Team began developing concepts for tools that could collect a variety of lunar samples as well as take the measurements needed to conduct geological studies. These designs were based on both the Sonett Report and the Falmouth conference report, with the latter providing some specific recommendations: a long-handled trowel (really a small shovel); a rock hammer; sampling tubes to be hammered into the lunar soil to collect small subsurface samples; a hand-held magnifying glass; a combination scriber and brush to mark and clean the samples; and sample bags and special sample containers, one of them airtight. A camera was also recommended. We began to build prototypes of these tools at

MSC and at Flagstaff, believing that eventually, regardless of whatever unique requirements we ultimately received from the still to be selected sample PIs, all these tools would be needed.

With the possible exception of the airtight container, these early tool and sample container lists constituted the standard inventory that any field geologist would recognize, modified for their unique application. Everyone knew, for example, what a geologist's hammer looked like. But some changes would be needed, since each tool would be used by a space-suited astronaut, perhaps under difficult lighting and temperature conditions, and in one-sixth gravity. We also had to factor in limited payload weight and stowage space, both on the trip to the Moon and returning. We knew that meeting all these constraints would require some compromises, clever design, and perhaps most important, careful input from the astronauts.

In September 1965, shortly after the Falmouth conference, Will Foster sent MSC a proposed second set of guidelines for Apollo science. In his memo he asked Robert Gilruth, MSC center director, and Max Faget to "prepare a Program Plan from which we can establish firm Program Guidelines to which all of us involved in this effort can work."[4] Foster's guidelines included discussions of sample return and lunar sample boxes, the Lunar Receiving Laboratory (LRL), the geophysical ground station, recently given the name Apollo Lunar Surface Experiments Package (ALSEP), and the geological hand tools and other equipment. He urged MSC to develop the guidelines as soon as possible, since we had little time to deliver the scientific equipment for the first missions.

While these guidelines were being developed we continued selecting the sample analysis PIs. After their proposals were received, Dick Wilmarth, Ed Chao, and Bob Bryson spent the next several months visiting the potential PIs and their labs to determine if they were equipped to conduct the analyses they proposed. Some were, some were not. As a result, OSSA began a program to upgrade the labs even though their proposals had not been officially approved. During the next five years, NASA transferred over $19 million to the sample PIs to purchase equipment and compensate them for their efforts.

As part of its responsibilities, the Field Geology Team began a careful review of the proposals by establishing a geology working group chaired by Shoemaker. In addition to Shoemaker, the working group consisted of Goddard, Mackin, and Waters from the Field Geology Team, Harry Hess (from the Space Science Board), and Ted Foss and Jack Schmitt from MSC. I served as secretary.

We met over a period of nine months, and at the end of 1966 we sent our report to OSSA. We recommended that almost all the proposals submitted be accepted, a total of forty-one.[5] At Dick Wilmarth's urging we also submitted a list of tests and experiments that should be conducted at the LRL, the equipment the lab should contain, and based on our ongoing studies, the types of containers that should be carried on the missions to hold the different types of samples we expected would be collected.

With Walter Cunningham immersed in his duties with Gemini and Apollo, our astronaut contact for the development of science equipment became Don Lind. Don had been selected in April 1966 as one of the nineteen astronauts in the fifth selection group, less than a year after the first scientist-astronaut selection. He had a Ph.D. in physics, and I had worked with him at Goddard Space Flight Center, where he was employed before his selection. He was an excellent choice to interact with the science community. Since he had also been a navy pilot and had a reputation at MSC as a meticulous worker, his opinions carried a lot of weight with the astronaut office. Jack Schmitt, as the only geologist-astronaut, would become closely involved in designing and developing the tools and experiments, but at this time he was just finishing his flight training.

Lind became our sounding board and made important contributions to Apollo science. He spent many hours trying each new design in a pressure suit, and along with Gordon Swann and other MSC and USGS staff he attempted to validate them in NASA's converted Air Force KC-135 (nicknamed the "Vomit Comet" for the reaction of many test subjects during the flight parabolas specially calculated to provide short periods of low or zero gravity). Ray Zedekar and others from the MSC Flight Crew Systems Division also worked tirelessly to test and improve the tools.

Simulations continued at Flagstaff through 1966 and 1967, prompting considerable refinement in the number and design of the hand tools the Field Geology Team would recommend. Astronaut mobility, dexterity, and visibility in the pressure suit were ultimately the major considerations and led to several unique tools not carried by geologists on Earth. In February 1967 a critical design review (CDR) of the Apollo lunar hand tools was held at MSC.[6] Because several of the proposed hand tools were not ready for the review, it was decided to designate a "hand tool pool." From the pool, a total of about twenty pounds of equipment could be selected for each mission, tailored to the mission's specific needs. A tentative priority list was established: tool carrier, sample bags

(100–200), maps, tongs, hammer, scoop, drive tube number 1, extension handle (used with several tools to eliminate bending over), gnomon, drive tube number 2, surveying staff (later dropped from the pool), color chart, drive tube number 3, sample bag dispenser and sealer, aseptic sampler, spring scale, and combination brush/scriber/hand lens.

The tool carrier, a three-legged stand, allowed the astronauts to carry their tools from station to station with one hand and then reach them without stooping. It was used on only two missions, *Apollos 12* and *14*. A second design carried on the J missions held the tools so that they could be mounted on the rear of the lunar rover.

The gnomon, a unique device, was devised by USGS to be placed in the field of view of the cameras the astronauts used on the lunar surface. It provided geometric and photometric control so that the photographs could be used to make analytical measurements. It consisted of a tripod about fourteen inches high supporting a gimbaled, weighted rod that would hang vertically. The shadow cast by the rod (hence gnomon) showed the direction the camera was pointed so that the astronaut need not estimate it and transmit it by voice. A gray scale on the rod was used for photometric calibration of the black-and-white photos, and a color chart on one leg helped us calibrate the color photos. With all this data available, we were eventually able to make stereo pairs from the photos and produce contour maps of the areas where the photos were taken.

The spring scale would weigh the rock boxes and individual sample bags brought back to Earth. These weights were important to the engineers doing trajectory analysis during the astronauts' return journey. Those who saw the movie *Apollo 13* may remember that Mission Control in Houston could not understand why the returning spacecraft did not respond as expected to the course corrections being made to bring the astronauts back within the narrow corridor in space required for a safe reentry. The combined LM and command module (CM) weights were accurately known, so they should have responded predictably to the small thruster burns. Finally someone remembered that the computer programs had been calculated allowing for a few hundred pounds of returned lunar samples. No samples were on board, since the astronauts had never landed on the Moon. When this figure was corrected and the proper weight inserted into the programs, the returning spacecraft was steered precisely into the Earth's atmosphere, allowing the command module to make a safe landing.

At this CDR, concerns again surfaced about the materials used in the tools. One dealt with the magnetometer experiment that would be deployed with the ALSEP and stowed near the tools on the LM. Stainless steel (the preferred material for the hammer and drive tubes, for example) might induce too much remnant magnetism, thus affecting the accuracy of its readings. Another concern was how hot or cold the tools would become in full sunlight or shadow, since the gloves used for extravehicular activity (EVA) could tolerate temperatures only in the range of $-250°F$ to $175°F$. It was decided that the tools would be anodized or given a gold tone to moderate temperatures on the surfaces the astronauts would touch.

Also at this CDR the surveying staff received a careful reexamination. To take full advantage of its capabilities the astronauts would have to make twelve settings at each station, taking a total of five to ten minutes. We were told the astronauts thought this was too long, and most of us agreed; their time on the lunar surface would be our most precious resource. The staff was eventually dropped from the pool. By the time the J missions flew, the "hand tool pool" was no longer required because the science payload was large enough to accommodate all the needed tools, some of which were new to the J missions or had been redesigned by that time.

With this background, we can now turn to sampling. The geology training the astronauts endured had one primary focus: to instruct them on what samples to collect and how to collect them. The training emphasized thorough verbal descriptions and proper photographic techniques to ensure good documentation of the sampling site. Sampling for geological analysis on Earth has progressed to a fine art, using techniques to fit the problem under study. Probably the greatest change in the past thirty years is the enormous amount of information we can now wring from a small sample (a few ounces or grams). Many of the types of analyses that let us extract this information from such small samples were in their infancy when we began planning for lunar sampling. But we knew that any samples brought back to Earth, no matter how small or large, would exponentially increase our knowledge of the Moon and its history. As we began to look closely at the issue and to assess the opportunities the Apollo landings would provide as well as their limitations, the sampling program became more and more sophisticated. This sophistication found its way into the types of samples wanted, the special tools needed to collect them, and the packaging or containment requirements.

Our first concern was the "grab sample" (later named contingency sample), one astronaut's first order of business once he was on the lunar surface. Everyone agreed on the importance of collecting this sample in case the first EVA was curtailed, but there was little agreement on how much should be collected, how and where it would be collected, how it would be documented, what tool(s) would be used, how it would be packaged (at one point someone suggested using a spare urine bag), where it would be stowed in the LM and the command and service module (CSM), and on and on. We first thought this sample should be passed back to the astronaut in the LM to ensure that something would be returned regardless of the outcome of the landing. This operation would mean using a significant part of the first EVA time to collect the contingency sample. These concerns held not only for the first landing but for all subsequent landings as well. In September 1967, after a review of the preliminary timelines at MSC, I raised these issues with Mueller's office, urging that they be addressed as soon as possible so we could proceed with tool and sample container design, which would in turn affect astronaut training and schedule development.[7]

Our next concern was the design of the large containers that would hold the samples on the return to Earth. They would have to be stowed in the LM on the outbound passage, then transferred to the CM for the return. Finding stowage space limited their size and weight and also their location relative to the spacecraft's center of gravity, since their weight would differ outbound and during landing maneuvers, during LM takeoff and on the CSM's return from the Moon. Heavy aluminum boxes, called Apollo lunar sample return containers (ALSRCs), or "rock boxes," were finally selected to satisfy these constraints.[8] They were designed and manufactured by Union Carbide at the Atomic Energy Commission's Y-12 plant at Oak Ridge, Tennessee. Each box weighed thirteen pounds and had an inner volume of less than one cubic foot, with outside dimensions of approximately $19 \times 11 \times 8$ inches. They were designed to withstand fifty gs and to maintain a vacuum seal in case of a hard landing in the ocean. Depending on the type of samples collected, each box could hold twenty to forty pounds of material. Two boxes would be carried on each mission, and after the samples were placed inside they could be sealed while on the lunar surface. The contract with Union Carbide called for the manufacture of twelve items of flight equipment and nine test containers. Two more flight containers were added later to the contract. When the boxes were opened at the LRL, high vacuums were always found, relieving some of the worry on the first three missions that alien organisms might have escaped into the Earth's atmosphere.

For collecting the contingency sample, a special tool was made with a long handle and attached bag. After the bag was filled, the handle would be disconnected and the bag placed in an astronaut's pocket in case they had to make a quick departure (thus resolving the question of spending time to get it back into the LM). With this limitation, small contingency samples were collected on each mission, always close to the LM, without much regard for the location, and not always documented with a photograph. After the contingency sample was safely in the astronaut's pocket, subsequent sampling became much more exacting. Depending on the mission and the prescribed timeline, further sampling might be postponed until later in the first and subsequent EVAs. This later sampling would be carefully planned to ensure that the landing site was covered as completely as possible within the radius of operations.

Another concern was what type of contamination would be introduced to the samples during landing by the LM descent stage engine exhaust. The exhaust, plus the astronauts' activities once they exited the LM, might introduce carbon compounds, making it hard to tell if any form of life existed on the Moon. In the summer of 1965 MSC gave Grumman (the LM manufacturer) and Arthur D. Little a small contract to study these questions. In November they briefed us on what they had determined.[9] There would, of course, be some contamination, estimated to be as much as one ton of various compounds spread over the landing site if they were all absorbed on the lunar surface. But chemical reactions could be predicted based on educated guesses about the composition of lunar soil, and they thought the contaminant molecules introduced by the exhaust could be identified during analysis of the lunar samples back on Earth. This study satisfied some, but not everyone, that the problem was understood, in particular the question of contamination from the astronauts' space suits.

Concern that the samples returned might harbor some unknown disease, and the opposite fear that the astronauts might contaminate the samples on the Moon, led to the development of a sampling device called the aseptic sampler. Its function was to retrieve a small sample from an area away from the landing site, where there would be a minimum chance that the exhaust from the LM descent engine would have introduced foreign material into the soil. The aseptic sampler was also designed and built by Union Carbide at the Y-12 plant, to specifications dictated by the National Academy of Sciences report on back-contamination. Its design became rather complicated. An extension handle would place a small coring tube against the surface a few feet from the "dirty" astronaut in his pressure suit. Two extendable feet would be unfolded to steady the sampler, and the astronaut would then pull a wire to open the coring device and push it into the soil. Surrounding the lower part of the handle was a sterile plastic bag into which the small core tube would be retracted; then the bag would be sealed to avoid any contamination after collection. All these functions were designed to avoid any contact with the astronauts or their gloves, because back on Earth the sample would be studied to detect organic compounds at a level of a few parts per million.

Dick Green, the ALSEP engineer and an office colleague, recalls being present at the final aseptic sampler training rehearsal by the *Apollo 11* astronauts. Sam Phillips was also there to witness the demonstration of another late addition to the astronauts' workload, a sore point with NASA management (which undoubtedly prompted Phillips's attendance). As might be expected, the complicated device malfunctioned. Phillips made an instant management decision to remove it from the flight and said contamination concerns would have to be resolved by studying the other returned samples (they were).

For the *Apollo 12* mission and subsequent ones, two new types of samples somewhat satisfied the requirements addressed by the aseptic sample: the special environmental sample and the gas analysis sample. But there was no attempt to isolate these samples as carefully as if the aseptic sampler had performed successfully. The special environmental sample was a small container, large enough to insert a drive tube; it was taken to the Moon tightly sealed to prevent any contamination during the outbound trip. Once a drive tube sample was retrieved on the lunar surface, the container would be opened, the tube inserted, and the container carefully resealed. The gas analysis sample was designed to obtain an uncontaminated sample of any constituents of the tenuous lunar atmosphere. The container was vacuum sealed on Earth and opened only after the astronauts were on the lunar surface. It would remain open for one or more EVAs, have a small amount of soil added, then be resealed, in hopes of capturing a few atoms or molecules that might be present in the near vacuum on the Moon.

To accommodate the procedures called for by the Field Geology Team and other scientists, several types of sample bags were designed. They would be modified as we learned from the experience of the astronauts using them on the lunar surface and the teams handling the samples back on Earth. In addition to the small Teflon bag that held the contingency sample, three other types of Teflon bags were designed to hold samples designated selected sample, documented sample, and tote bag sample.

The bags for the selected sample (which replaced the bulk sample collected on *Apollo 11*) could contain a large volume of sample and have enough space to store the core tubes plus the lunar environment and gas analysis samples. The smaller documented sample bags (seven and a half by eight inches) were carried on a twenty-bag dispenser and would be removed individually to hold samples documented by the astronauts' description and photographs. Each bag was premarked with an identification number that would be relayed back to MSC as the bag was filled to obviate confusion when the sample was opened at the LRL. After the selected and documented sample bags were sealed, they were placed in the ALSRCs. The large tote bag would hold any large rocks the astronauts collected. This bag would not be placed in an ALSRC but would be separately stowed, first in the LM and then in the CM.

Cameras had been part of the astronauts' equipment since the first Mercury flights. From Gemini flight GT-4 on, they were included in formal experiments. Some good science had resulted from the pictures of Earth taken during the Gemini flights, especially new views of important terrestrial features such as the Himalayas and impact craters never before photographed.[10] Cameras would become an essential element in each Apollo mission to preserve what the astronauts saw on the lunar surface and in lunar orbit.

On the Moon, cameras were needed for three purposes: to document the individual samples collected; to provide detailed views of the areas where the astronauts were working as well as panoramic views; and to record the placement of the ALSEP central station and experiments and of any other experiments the astronauts deployed. The Hasselblad camera, which all the astronauts were used to and which was already qualified for space flight, was an immediate candidate for lunar surface photography. Other types of cameras would be added in the months ahead, but the Hasselblad soon became the top choice.

Shoemaker and his Field Geology Team also believed that stereoscopic photographs were necessary to document samples and the general geological scene.

He enlisted Homer Newell, who agreed and wrote to George Mueller that they were "a necessity on every lunar landing mission."[11] In the summer of 1966 the Manned Space Flight Experiments Board asked Shoemaker to develop the specifications for a stereo camera. Preliminary work was carried out to develop such a camera, but it was eventually canceled because of payload weight and EVA time constraints. The astronauts were then trained to use the Hasselblads to take stereo pairs.

Integrating the cameras with the astronauts' activities became a major challenge. They had to be handy but not in the way. How would the astronauts carry, point, and trigger them in their space suits and clumsy gloves? After many trials and errors, the solution was to mount the cameras on the astronauts' remote control unit, a fixture attached at chest level on the outside of the pressure suit. A dovetail bracket on the remote control unit allowed the astronauts to slip the cameras on or off with some ease. Test subjects and the astronauts soon became adept at pointing the cameras and compensating for the parallax caused by the camera's being below their line of sight. Camera controls were modified to be used with gloves. Once this camera was accepted, most of the simulations and training sessions included the Hasselblads, to determine how best to document the projected lunar surface activities and to get the astronauts used to them.

The camera inventory carried in the LM for use on the lunar surface was extensive. One television camera, three 70 mm Hasselblads (two with 60 mm lenses and one with a 500 mm lens), one 16 mm Mauer sequence camera mounted in the LM pilot's window (to photograph the landing, initial surface activities at the foot of the LM ladder, and rendezvous maneuvers with the CSM in lunar orbit; it was used in later missions on the lunar surface), and about twenty-five film magazines of various types. A seventh camera, the Apollo lunar surface close-up camera (ALSCC), was one of the late additions to the science equipment.

The ALSCC was Tommy Gold's last attempt to reap some fame from the lunar landings. Still obsessed with the nature of the fine material that constituted the lunar soil, he proposed a special camera to take close-up stereoscopic photographs of it. He submitted a proposal in 1968, and after some debate on its merits, the SSC finally agreed to carry his camera. Shoemaker and the Field Geology Team were incensed at this decision, believing it had little scientific merit and, most important, would take time on *Apollo 11* and the next missions from the much more important geological tasks and the sampling. Our office supported Shoemaker's reasoning. We also knew that we would be assigned to oversee the rapid development of the camera while dealing with a potentially difficult PI. We were overruled, and the camera development went forward.

Gold's photographic objectives required a complicated design for an entirely new type of camera. He wanted the camera's focal plane to be very short, in lieu of magnifying lenses, so that particles of 0.1 mm or even smaller could be distinguished and measured; achieving this called for taking stereoscopic pairs with the camera close to the lunar surface. Since the astronauts could not bend low enough to set a camera on the surface and operate it, the camera would have to be attached to a long handle. With the camera essentially in direct contact with the surface, a light source would also have to be provided to flash for each stereo pair. On and on went the design requirements for this strange contraption that few favored, including the astronauts, who were vocal in their objections to using it. So much for the politics of science—Tommy had friends in high places.

To add to the complications, when the NASA Procurement Office learned of our plans to get bids to design and build the camera they insisted it be made a "small business" contract. The government's policy of giving contracts to small businesses deserves support, and my government career after I left NASA depended on small business for its success, but this was a bad decision that we knew would give us trouble. Schedules were tight, and the camera's design would require some clever engineering. We scrambled around and finally located a company (its name escapes me), and MSC awarded a contract. Robert Jones at MSC was named program manager. After several months of monitoring the company's progress, it became clear that it would be unable to deliver the camera on schedule, if ever.

Now we were in real trouble, since the camera was scheduled to be carried on the first landing mission and we had lost almost six months. But because of the tight schedule, in January 1969 we were able to justify awarding a sole-source contract to the most qualified supplier, Eastman Kodak. Kodak worked literally around the clock and delivered the flight hardware and training cameras on schedule to meet the *Apollo 11* launch date. Gold's camera performed almost flawlessly, thanks to the Kodak engineers, and it was also carried on *Apollo 12* and *Apollo 14*. Although it was not a favorite experiment for the astronauts—a few threatened to throw the camera away—they complied with most of his requests for his unusual photographic subjects and returned forty-nine and a half stereo pairs.

How much new science resulted from analysis of the photographs is debatable. Gold tried to use them to advance some of his pet theories, and David Carrier, an MSC engineer who had provided oversight on the soil mechanics experiment, reminded me that when he and several other MSC staffers cooperated with Gold in writing his report for *Apollo 14* they withdrew their names as coauthors because they disagreed with some of his conclusions.

When more weight became available on the J missions, the tool inventory remained essentially the same except that we added a rake, suggested by Lee Silver after the *Apollo 12* mission when the astronauts found it difficult to pick up small rocks and collect samples mixed with the lunar soil. We reasoned that such samples would yield a wide variety of lunar rocks, since every landing site might contain ejecta from many distant sources. The rake was designed as a scoop, closed at one end, with wire tines spaced about a quarter inch apart to sift out the loose material but retain the larger pieces. It was used successfully on all three J missions.

We added another important piece of equipment for the J missions, the Apollo lunar surface drill. Two requirements led to its development: the ALSEP heat flow experiment, which needed two holes for inserting the sensors, and the geologists' and geophysicists' desire to obtain subsurface samples. Here once again the experience gained in studying a deep drill for the post-Apollo missions was valuable. Jack Hanley, detailed to my office from USGS, had monitored the hundred-foot-drill studies at Marshall Space Flight Center, and he was assigned to oversee the drill. The RFP released by MSC called for bids to build a drill that would extract cores to a depth of one hundred inches. The competition was won by Martin Marietta, Denver, teamed with Black and Decker.

The design the Martin Marietta team selected was a battery-powered rotary percussive drill in which the power head imparted short impacts at the same time as the drill pipe (core stem) rotated. The astronaut could also lean on the drill handle to add force and improve the penetration rate. The core stems (a total of six that would be screwed together during the drilling) were fluted on the outside, as in the hundred-foot drill studied by Westinghouse several years earlier, to carry the cuttings or soil to the surface as the drill penetrated into the subsurface. Each core stem, made of fiberglass tubular sections reinforced with boron filaments, was about sixteen inches long. As each one penetrated to its full length, the drill head would be disconnected and another core stem screwed on to continue drilling. A tripod device held the extra sections above the ground until they were connected during the drilling. There was enough battery power to drill three holes: two for the heat flow experiment and one for the core sample.

After five Surveyor spacecraft had landed on the Moon and returned pictures and rudimentary data on the characteristics of the lunar surface, many questions still remained about some of the engineering properties of the upper layers of the lunar surface. Since the Surveyor spacecraft had not disappeared in fluffy dust, we now knew that traveling on the lunar surface in some sort of wheeled vehicle would be possible. Using lunar soil to shield shelters while lunar bases were being built (as proposed in the Lunar Exploration System for Apollo studies) also appeared feasible, but more hard data were needed to understand how these soils could be excavated.

The need to predict the behavior of lunar soil, insofar as it would affect the design of vehicles and other equipment, as well as the need to collect other basic information, led to the inclusion of a soil mechanics investigation on the final four Apollo missions. This experiment, closely allied to the field geology studies, consisted of analyzing the astronauts' observations on the character of the soil as they moved about; photographing the soil after it was disturbed by their activities (e.g., boot prints, tire marks, and trenches), augmented by physical measurements made in situ with penetrometers and other devices; and finally, making measurements on the returned samples.

James Mitchell, from the University of California, Berkeley, was selected as the soil mechanics principal investigator. His team included as coinvestigators Nicholas Costes from MSFC, who had been on the *Apollo 11* and *Apollo 12* Field Geology Team and had participated in some of our post-Apollo studies, and Dave Carrier from MSC. Don Senich, a former instructor at the Colorado School of Mines who was detailed to my office from the United States Army Corps of Engineers, was to oversee the development of this experiment from headquarters.

A simple penetrometer, consisting of a long aluminum shaft slightly less than half an inch in diameter, was carried for the first time on *Apollo 14.* It was to be pushed into the surface at several places near the LM to a maximum depth of sixty-eight centimeters. Black and white stripes were painted on the shaft, and after pushing it as deep as possible each time, the astronaut would read back the number of stripes still above the surface as a measure of the depth achieved. Mitchell's team would then calculate the forces involved by applying data obtained from terrestrial simulations. On the *Apollo 15* and *Apollo 16* missions a more sophisticated, self-recording penetrometer was carried. This device consisted of a base plate, a shaft with two different-sized interchangeable nose cones, and an upper housing containing the recorder. An extension handle above the recorder helped the astronauts force the nose cones into the surface. After pushing the penetrometer into the soil, they would remove the data drum from the recorder and return it for analysis.

Chapters 11 and 12 will tell more about how the equipment for the field geology experiment was used on the Moon by the crews of the six landing missions.

The Apollo Lunar Surface Experiments Package and Associated Experiments

By 1964 the growing fraternity of space and lunar scientists began to see the Apollo missions as an opportunity to address many age-old questions. These questions related not only to the Moon itself but to the Earth, the entire solar system, and to some degree the whole universe. The Moon would provide the equivalent of a spacecraft on which to conduct experiments never before possible. The Sonett Report, along with advisory panels from the Office of Space Science and Applications, the Office of Manned Space Flight, and the National Academy of Sciences' Space Science Board, guided us in soliciting experiments to be associated with a permanent science station such as we studied for post-Apollo missions under contract at Marshall Space Flight Center (these studies became the basis for the Apollo Lunar Surface Experiments Package, or ALSEP, developed for Apollo missions). We also solicited additional experiments that could be conducted on the Moon's surface independent of a geophysical ground station. At this time a few of the scientists who were thinking about experiments on the Moon were also considering how to conduct experiments in lunar orbit. Aside from Lunar Orbiter, however, there were no "approved" plans to provide a platform for lunar orbit experiments in the Apollo missions. I emphasize "approved," for though planning for such experiments was going on, no specific spacecraft had been designated to carry them. Experiments as well as cameras had been solicited for the Lunar Orbiter program, but the proposals were on the shelf until a program was approved.

Will Foster's February 13, 1964, memorandum, in addition to recommending a Field Geology Team that would help plan for sample collection, listed four geophysics teams, selected to recommend and design lunar seismic, magnetic, heat flow, and gravity experiments.[1]

The seismology experiment was divided into two parts, passive and active (each requiring different instrumentation), to monitor the Moon's internal activity (moonquakes) and determine its shallow and deep structure. The team consisted of Frank Press, then at California Institute of Technology, and Maurice Ewing and George Sutton of Columbia University. The memo proposed additional investigators for the active experiment, but they were unnamed.

A third type of seismic experiment, engineering seismology, was also listed, to provide data that would be used for post-Apollo mission planning. Although considered a nonscientific experiment, it was designed to measure the Moon's surface characteristics to a depth of fifty feet. For this team Foster suggested personnel from the Manned Spacecraft Center and the United States Geological Survey at Flagstaff, since USGS had begun to study the data needed for designing lunar bases and mobility devices under my office's contract with Gene Shoemaker. The engineering seismology experiment was finally designated the active seismic experiment, and Robert Kovach at Stanford University became the principal investigator (PI), supported by coinvestigators Thomas Landers, also from Stanford, and Joel Watkins, who had moved from USGS at Flagstaff to the University of North Carolina. Kovach never selected anyone from MSC to join his team.

The magnetic measurements team consisted of James R. Balsley of Wesleyan University, Richard R. Doell from USGS, Norman Ness of NASA Goddard Space Flight Center, Chuck Sonett from NASA Ames Research Center, and Victor Vaquier from the University of California, San Diego. This team was to specify the magnetic measurements needed to determine the lunar magnetic field (if any) in the presence of solar and interplanetary magnetic fields and the methods for measuring any remnant magnetic field in lunar rocks. All previous attempts to measure a lunar magnetic field from a distance had failed to find any significant field; thus these measurements would be critical in unraveling the Moon's early history.

The heat flow measurement team consisted of Francis Birch from Harvard, Sydney P. Clark from Yale, Arthur H. Lachenbruch of USGS, Mark Langseth from Columbia, and Richard Von Herzen from the University of California, San Diego. In addition to designing the heat flow instrumentation, the PI for this experiment would become closely involved with the design of the Apollo lunar drill, because the heat flow sensors would be lowered into two holes made by the drill.

The final team listed in the memo was to make gravity measurements. It consisted of Gordon MacDonald from the University of California, Los Angeles, and Joseph Weber from the University of Maryland. This experiment, it was hoped, would provide some of the more exotic measurements to be made on the Moon; not only would it measure the deformation of the Moon created by the pull of the Earth's mass, but it might detect gravity waves, predicted by Einstein but never unequivocally measured. This experiment was truly unique to the Moon, since to have any hope of recording gravity waves the instrument, a sensitive gravimeter, had to be on an extremely quiet body, as many believed the Moon to be.

These teams, like the field geology, geochemistry, petrography, and mineralogy teams, were also approved by the Space Science Steering Committee (SSSC). My purpose for listing the team members is twofold. First, it shows for the record that their members included many of the leading scientists of the day in the identified fields. Thus this obviated the need to make the usual formal solicitation to the scientific community as a whole, since it would undoubtedly have resulted in teams similar to those proposed. Some might take issue, but I believe that is true, since only a few leading scientists in these fields were considering lunar experiments. This procedure shortened the time it took to get the key players in place, probably by six months or more—not an insignificant consideration. Second, the makeup of the teams changed with time, especially the important position of PI for each experiment. This position, of course, was the key to future scientific fame, for the PI's name would appear first in the final reports and citations.

Each team was to design and build its experiment through the prototype stage, training the astronauts in its use or deployment and, finally, reducing and reporting on the data returned. This opportunity was extremely important, because Apollo promised long-term data collection for experiments attached to the lunar ground station (one year or perhaps longer) and exciting data transmission (bandwidth) capability. Weight and power allowances were expected to be generous compared with a typical unmanned spacecraft experiment, and having an astronaut set up the experiment or return some or all of the data would add to the value. Some people in the unmanned science camp argued that using the astronauts for those types of experiments was an unnecessary complication, but in general their involvement was considered a real plus. Before this date in 1964 we had little experience deploying unmanned payloads either in space or on the Moon, and those that had been deployed were relatively unsophisticated. Using astronauts to set up or operate an experiment had only occasionally been factored into an experiment's design for the Gemini program, so this would be a new challenge to the scientific community.

After SSSC approved Foster's recommendations, contracts were negotiated with the team members, and OSSA began to fund and manage their efforts. As promised in Foster's memo, other experiments and investigators were brought on board later to cover important areas of science not included in the initial teams. Experiments added during this time were the Solar Wind Spectrometer (SWS) to measure the solar plasma striking the lunar surface; the Suprathermal Ion Detector Experiment (SIDE), which could measure a variety of interactions with the solar wind and complement the SWS measurements; and the Cold Cathode Gauge (CCG) to measure the composition of the lunar atmosphere. These experiments would also be attached to the ground station for their power, housekeeping, and data transmission needs.

Another experiment that would operate independently of any ground station, the Solar Wind Composition experiment, was also approved for the first missions. It was proposed by a Swiss team headed by Johannes Geiss from the University of Bern. Its purpose was to collect and return solar wind ions, such as helium and neon, to help us understand the composition of the solar wind. This experiment was funded in part by the Swiss government.

With the initial suite of experiments and experimenters under contract, in early 1965 our efforts turned to the design and development of the station that would support the experiments. By this time the MSFC Emplaced Scientific Station (ESS) contractors, Bendix and Westinghouse, had progressed to a preliminary design of a geophysical station for post-Apollo missions. It had all the characteristics we wanted for an Apollo station; the major differences were in overall size, the ESS being larger than we could expect for the first Apollo landings.

On May 10, 1965, Foster sent Ernst Stuhlinger at MSFC a request to submit a work statement for an Apollo scientific station.[2] At the same time he also asked Max Faget at MSC to submit a similar work statement. Much to our chagrin, George Mueller's office, led by James Turnock and Leonard Reiffel, thought MSC should be the lead center in managing this complex payload. I was lobbying hard for MSFC and had convinced E. Z. Gray and Will Foster that, based on all the work MSFC had done for our post-Apollo mission studies, it was the best qualified. MSC had nowhere near as much experience with lunar science payloads, and it lacked qualified staff to oversee such a contract.

This controversy came to a head at a Saturday meeting with George Mueller, on May 24.[3] (Remember the best day to get his undivided attention?) Also at the meeting (besides me) were Sam Phillips, Edgar Cortright (Mueller's deputy), E. Z. Gray, Will Foster, Dick Allenby, Jim Turnock, Len Reiffel, Benjamin Milwitzky, and Jack Trombka. The major issue was deciding who would manage the Apollo science station. We reviewed the two work statement proposals from MSC and MSFC and weighed the strengths and weaknesses of each. We described the problems we had working with MSC on matters dealing with science and the much better relationship we, and the scientists we had brought into the post-Apollo studies, had with MSFC.

After about two hours of discussing the pros and cons, the MSFC work statement was judged superior to MSC's and likely to elicit the best proposals. I thought we had carried the day and that MSFC would be assigned this important task. But Phillips finally weighed in with his opinion—that in spite of its deficiencies MSC should become NASA's "lunar expert." Mueller agreed and also expressed his unwillingness to have Stuhlinger manage the Apollo science program. Why he was uncomfortable with Stuhlinger he never explained. He did agree that the MSFC work statement should be the basis of the request for proposals and asked that three companies be selected in the initial study to ensure some competition.

The anointing of MSC as our lunar expert was a devastating blow to many of us in attendance and presaged the at times bitter disagreements we and the PIs would have with the MSC managers in the years ahead. As a gesture to assuage our disappointment, Mueller asked us to prepare a history of our past year's negotiations with MSC so that he could understand the situation. Perhaps Mueller's review of our tales of past disagreements was a factor in the decision to transfer Bill Hess to MSC at the end of 1966 to lead the science activities there.

The MSFC's Apollo work statement, based on the ESS study, in essence called for a junior ESS. Because extended periods of data collection were needed for many of the experiments selected, it was decided that the station would be powered by a radioisotope thermoelectric generator, the same power source proposed for the ESS. RTGs, already under development for planetary spacecraft that would fly too far from the sun to collect sufficient solar energy, generate electric power through the decay of radioactive elements, in this case plutonium. This decay produces heat, which is in turn converted by thermocouples to electric power. RTGs were an ideal power source for lunar-based experiments, because for fourteen consecutive Earth days of every day/night lunar month cycle, the station would be in darkness and very cold. Batteries alone could not do the job; they would be far too heavy to accommodate the required duty cycles. A solar-powered station would have required a large solar array, would be difficult to deploy on the lunar surface, and would still require a large, heavy battery pack to sustain it during the fourteen-day lunar night. When the solar array and the batteries were studied, it became evident that RTGs provided not only a distinct savings in weight but also greater reliability and simplicity, because among their other advantages they have no moving parts.

The Atomic Energy Commission (AEC), the agency responsible for overseeing the manufacture of RTGs, had several well-tested designs to choose from that could provide various amounts of power depending on how much weight one could allocate to the power source. The RTGs were manufactured by General Electric's Valley Forge, Pennsylvania, plant, with the plutonium supplied by the AEC. Safety considerations were the primary arguments against using an RTG. First was the question of how an astronaut could safely deploy the RTG. It would be "hot," both in temperature and in radioactivity. Second was the chance of a mission abort in which the plutonium 238 fuel source might be released into the atmosphere. Plutonium is highly toxic if inhaled. AEC and General Electric believed they could solve both problems, and later ground handling tests and destructive tests exposing the system to high-energy impact and heat loads proved them correct.

The RTG power source (system for nuclear auxiliary power, SNAP) selected to provide power for a year or longer was designated SNAP-27. It consisted of a fuel capsule and generator. The fuel capsule would be carried to the Moon on the lunar module descent stage in a special graphite container, and after the astronauts removed it and inserted it in the generator, it would provide 63.5 watts of electrical power to the central station. With a fuel half-life of almost ninety years, it more than filled the need for a long-term power source.

(The RTG on the *Apollo 13* mission, still attached to the LM lifeboat that sustained the astronauts during that harrowing, nearly fatal experience, reentered at a speed far beyond that anticipated for a typical Earth orbit failure, but it is believed to have survived intact, as designed. If the cask protecting the plutonium heat element had failed, the sensitive instruments on the aircraft sent to sample the air at the reentry point over the Pacific Ocean would have detected plutonium in the atmosphere.)

Once the power source had been decided, the next critical step was selecting a design for the communication and data relay subsystem. Commands would be transmitted from Earth to control the station and its experiments, and data would be relayed back from the lunar surface. NASA's Manned Space Flight Network (later incorporated into the Space Tracking and Data Network) would provide round-the-clock monitoring, eliminating the need to provide data storage at the station as we had envisioned for some of the ESS experiments. Raw or processed data would then be given to the PIs for reduction and interpretation. A difficult question was, How much data should the station be capable of handling? No matter how much was made available, PIs would always be hungry for more. Until specific instruments were designed, this would remain an open question. At the Falmouth conference, attended by some of the probable experiment PIs, it was recommended that the station be designed to accept various types of experiments so that the instrument complement could be easily changed, depending on the landing site and the experiments required to answer site-specific questions. All in all, it would be a tough design challenge, but based on the work we had done for our post-Apollo studies, we felt confident it could be met.

In June, using the MSFC work statement as its model, MSC released the request for proposal (RFP) for the geophysical ground station that came to be known as the Apollo Lunar Surface Experiments Package. Max Faget's Engineering and Development Directorate's new Experiments Program Office was named MSC program manager, reporting to Robert Piland, recently appointed to head the office. Nine companies submitted proposals, and as Mueller had requested, three were selected to provide a preliminary design. In August, Bendix Systems Division, Space General Corporation, and the TRW Systems Group were each awarded a contract for $500,000. Each company would provide a preliminary design and mock-ups, to be delivered to MSC and Grumman at the end of the six months.[4]

The RFP requested that each design include a seismometer, heat flow sensors, magnetometer, a suite of atmospheric and radiation sensors, and a device to measure the micrometeorite flux at the lunar surface. (This last device, proposed by MSC and rejected by the Planetology Subcommittee four months earlier as not relevant to lunar science, had found its way back into consideration. MSC used its position as NASA's "lunar expert" to push one of its pet ideas.) The weight allowance for the entire package was not to exceed 150 pounds. After a review and evaluation of each contractor's design and performance during the six months, we planned to select one contractor to provide the flight hardware.

After the mock-ups were delivered, we convened a selection committee to decide which of the three teams would build the flight hardware. Bendix had obviously profited from its part in our post-Apollo studies of the ESS. Its preliminary ALSEP design was judged the most responsive to our requirements, and a contract was awarded in March 1966. With an initial value of $17 million, the contract finally grew to $58 million through increases in the number of flight and test units required and the added job of building four ALSEP experiments for the PIs and integrating more experiments than originally projected.

The contract and its subsequent modifications called for the manufacture of six flight-qualified ALSEPs, a "dummy" unit to fly to the Moon in the storage bay of the *Apollo 10* lunar module, prototype and qualification units, two training units, and one unit dubbed the "shop queen," which was modified and cannibalized and was generally available to test ideas. Joseph Clayton, later promoted to division general manager, was the initial Bendix program manager and was succeeded by Chuck Weatherred at the time ALSEP progressed to the prototype phase. Chuck, who had been closely involved with many of our post-Apollo studies, then continued as program manager through the missions.

Some additional details now about ALSEP, the attached experiments, and the other experiments that were deployed at the landing sites but were not dependent on ALSEP for their operation. First the ALSEP central station. The central station was the control center for the many instruments that were so carefully crafted by the experiment teams, some designed to record sensitivity levels unachievable for similar Earth measurements. The central station data subsystem would receive and decode the uplink commands for each experiment and collect the scientific and housekeeping data and transmit them back to Earth. A small helical S-band antenna would be mounted on top of the station and pointed by the astronauts to provide the data link to Earth.

Most of the experimenters were interested in collecting data over a long period, in most cases the longer the better. The ALSEP design goal was to survive for a minimum of one year, providing power, housekeeping functions, data collection, and transmission. This was no mean task, given the extreme temperature fluctuations (over 500°F) experienced on the Moon every twenty-eight days. At the same time that instruments or devices would be experiencing these temperature changes, they and the central station would be operating in a high vacuum. Lacking the normal methods of regulating experiment temperatures on Earth, their design would have to include novel ways to both heat and cool all the components.

Keeping the experiments warm was not as hard as keeping them cool; heat could be supplied by small electrical heating elements of various designs. But since liquid coolants could not be used, radiators, thermal blankets, and shielding were employed, utilizing new materials. In addition, the central station and the experiments would have to be carefully oriented to provide selective shadowing and reflection of the sun's rays.

Thirteen experiments were ultimately selected to operate with the five ALSEPs deployed on the Moon. (Some were on the ALSEP carried on the *Apollo 13* mission, and their remains are at the bottom of the Pacific Ocean.) Each ALSEP had a unique combination of experiments, ranging from four to seven, and some experiments were carried several times. The eight listed at the beginning of this chapter were considered of highest priority. Four more would be added over the next few years, plus a dust detector to help monitor ALSEP's health if dust or dirt on thermal surfaces caused a problem.

One of the four new experiments, Seismic Profiling, had an objective similar to the active seismic experiment but would provide additional information on the Moon's shallow structure. The other three were the Lunar Ejecta and Meteorites Experiment to measure the direction of travel, speed, and mass of micrometeorites arriving at the lunar surface (not the MSC proposal mentioned earlier); the Charged Particle Lunar Environment Experiment for measuring a wide range of charged particles caused by the interaction of the solar wind on the lunar surface; and the Lunar Atmosphere Composition Experiment, a spectrometer that would measure the composition and density of whatever gas molecules might be found in the tenuous lunar atmosphere. Some of the experimenters did their own contracting and built their experiments, delivering them to Bendix for integration, and some used Bendix as their contractor.

Nine other experiments, not dependent on ALSEP and not including those discussed in chapter 5, were to be deployed by the astronauts either in the vicinity of the LM or during their traverses. They fell into two categories: those used for studying the Moon and those that used the Moon as a convenient platform from which to make measurements.

As I mentioned earlier, one could think of the Moon as a spacecraft circling the Sun, on which you could place instruments to measure phenomena occurring within or outside our solar system. In our post-Apollo studies we had examined using the Moon as a site for astronomical observations, and this preliminary study elicited some interest from the astronomy community during the Falmouth and Santa Cruz conferences.

On later missions, when larger payloads became available, we had the opportunity to test this idea. An ultraviolet camera-spectrograph was proposed and carried on *Apollo 16*, the second J mission. The objective of the experiment was to evaluate the Moon as an astronomy base and to take pictures of targets in the far ultraviolet portion of the electromagnetic spectrum, a frequency that could not be studied from the Earth's surface because of our intervening atmosphere. The experiment was proposed by George Carruthers of the Naval Research Laboratory, and the instrument was designed and fabricated at his lab.

A second experiment in the category of using the Moon as an observation post was the Cosmic Ray Experiment, a multipart experiment proposed by three teams, one at the General Electric Research and Development Center, a second at the University of California at Berkeley, and a third led by Caltech. Its objective was to detect high and low energy particles emanating from the Sun and from outside the solar system. It had the potential to record particles that had not been detected on Earth, again because of interactions with our protective magnetic fields and atmosphere. It would go to the Moon mounted on the LM descent stage, where it would be exposed just after translunar injection, then folded and retrieved at the end of the third EVA. A related part of the experiment was a detector carried inside the astronauts' helmets to determine their exposure to cosmic rays during their transit and stay on the Moon or while in lunar orbit. This was important information because it concerned the astronauts' health, especially if a solar flare or some other major event that occurred somewhere in the universe at an earlier date would expose them to high energy particles during the mission. It was also important for planning longer-duration, manned missions to Mars.

Five of the nine experiments fell into the category of studying the Moon through various measurements. These were the Lunar Neutron Probe, the Laser Ranging Retro-Reflector (LRRR), the Lunar Portable Magnetometer, the Lunar Traverse Gravimeter, and the Surface Electrical Properties (SEP) experiments. By the time the last three were proposed, it was known that a small vehicle would be available to the astronauts, so the magnetometer, gravimeter, and SEP were designed to be carried on the lunar roving vehicle (LRV), with measurements taken either at the astronauts' discretion or at planned points. The magnetometer and gravimeter would measure the Moon's gravity and magnetic fields to determine if these values changed as the astronauts moved away from the LM. The SEP used radio waves to penetrate the lunar surface to look for layering in the Moon's crust. If there was no moisture in the upper layers, it might be able to penetrate deeper than the Seismic Profiling experiment. If water or ice occurred below the surface, the signals received would reveal their presence. The neutron probe would be lowered into the drill hole after the core was extracted to measure the rate of neutron capture below the lunar surface. This measurement would help us understand the physical processes that produced the lunar soil. After remaining in the drill hole for some time, the probe would be recovered and brought back to Earth for study.

The LRRR was a late addition to the roster of Apollo experiments and deserves further description. A laser beam, originating at an observatory on Earth, would be reflected from the Moon back to the observatory and thus provide an accurate determination of the Earth-Moon distance (within a few inches). It was proposed by Carl Alley from the University of Maryland and was built in time to be carried on *Apollo 11*. Alley was supported by a host of coinvestigators; one of them, James Faller from Wesleyan University, became the PI for the *Apollo 14* and *Apollo 15* missions. The experiment was designed and developed by the A. D. Little Company of Cambridge, Massachusetts, and built by Bendix. The experiment carried on the *Apollo 11* and *Apollo 14* missions consisted of one hundred reflectors, each about an inch and a half in diameter, arranged in a ten by ten square. They were mounted on an adjustable support that could be tilted and aimed at the appropriate angle for each landing site to best reflect the laser beams coming from Earth. The astronauts aimed the device using a sun compass and a bubble level, pointing the array at the center of the Earth. Individual corner-cube reflectors were manufactured under a separate contract by the Perkin Elmer Company. Because of difficulties in locating the LRRR at Tranquility Base and the Fra Mauro landing sites, the array carried on *Apollo 15* was increased to three hundred reflectors, and it proved much easier to locate and reflect laser beams from Earth.

A network of three LRRRs was to be placed on the Moon, separated as far as possible in latitude and longitude. By sending laser beams from the Earth to the LRRRs and bouncing them back, it was anticipated that the Earth-Moon distance could be calculated within approximately six centimeters. Such precise measurements would permit the study of many physical properties of the Earth and the Moon, including fluctuations in the Earth's rotation rate, the wobble about its axis, the shape of the Moon's orbit, and the Moon's wobble about its axis. Ultimately, if enough stations on Earth were capable of sending laser beams to the Moon, small movements in the Earth's crust might be measured. (Crustal movements are no longer measured this way. Instead, accurate laser ranging measurements are made from Earth to orbiting satellites. The LRRR, however, is the only experiment carried to the Moon by the Apollo astronauts that is still used for other types of measurements.)

Headquarters management of ALSEP was initially under the direction of OSSA in the Lunar and Planetary Programs office managed by Oran Nicks, and OSSA provided the funds to get ALSEP started. (The vast majority of ALSEP funding eventually came from OMSF.) William "O. B." O'Bryant, a retired navy captain, was named program manager, and Dick Green, a retired air force officer returning to NASA after a recall, was named program engineer. Ed Davin, still reporting to Will Foster, was named program scientist. I also maintained an oversight of ALSEP because of its close relationship to other programs I was managing, such as the lunar drill. Relations between headquarters staffers and MSC soured almost immediately. MSC continued its practice of not notifying headquarters when important reviews were to be held at Bendix or MSC. This caused a great deal of heartburn at headquarters. This attitude and way of doing business eventually led to the appointment of John "Jack" Small, who proved easier to work with; but the atmosphere had already deteriorated, and an uneasy relationship continued even when Small was replaced toward the end of the program by Donald Wiseman. Fortunately from our perspective, Bendix proved to be a cooperative contractor and recognized the importance of maintaining good relations with headquarters. This was a wise move, for in the ensuing six years there were a number of times when obstacles and difficult decisions arose that required the intervention of headquarters.

In some small defense of MSC's reluctance to keep headquarters apprised of ALSEP's progress, a careful line was always drawn between NASA's contract offices and its contractors in order to avoid any misunderstanding about who was in charge. MSC had the sole authority to control the ALSEP contractor's actions, and any changes to the contract scope could occur only with written direction from the MSC program manager. Probably all NASA centers had experienced instances when a contractor had used a conversation with someone from NASA headquarters to attempt to modify the scope of its contract, a surefire way to screw up the contract and make the center in charge see red. O.B., with his navy background, was a no-nonsense manager, and never to my knowledge did he create this kind of problem. But he never backed down from exercising his management prerogatives, which included the right to suggest changes to the program manager if he or his staff saw trouble developing and to keep a tight rein on the funding. O.B., Green, and Davin also felt that they were often the only ones sticking up for the experimenters when trade-offs were required, and they didn't hesitate to make their concerns known.

Toward the end of summer 1968, with ALSEP development in its final stages, NASA management began to reevaluate the first landing mission's lunar surface activities. Concern was growing about how well the astronauts would function on the Moon and, more important, how the LM would perform. Several ways to alleviate these concerns were explored. First the number and length of EVAs could be reduced. But if only one EVA was allowed, then ALSEP could not be deployed and still leave time for the astronauts to carry out their other important tasks, including sample collection. Not carrying ALSEP would reduce the astronauts' workload and the weight of the LM for the first mission—a partial solution to both concerns. Removing weight would also add a few seconds of hover time. ALSEP became a prime target for removal.

When rumors spread that the scientific experiments on the first landing would be drastically reduced, Charles Townes, chairman of the Science and Technology Advisory Committee, went to NASA senior management and argued for keeping as much science as possible on the first mission. Our office, Bill Hess at MSC, and others in the scientific community were also lobbying hard to keep ALSEP on the first landing mission and to maintain two EVAs. Our office was fighting for more than just the science. ALSEP and the geological tasks the astronauts were scheduled to carry out represented years of planning and hard work, not to mention suffering through many a contentious meeting with those in NASA who did not embrace the need to include science on Apollo. We were not prepared to accept such a defeat.

In September and October, in response to this outcry, our office and MSC studied an alternative to dropping the full ALSEP and presented it to the NASA Senior Management Council. A new, much smaller, and more easily deployed science payload was proposed and approved and given the name Early Apollo Experiments Scientific Package (EASEP). EASEP would comprise just three experiments, the passive seismometer, packaged with the dust detector, and the LRRR. Another self-contained, easily deployed experiment, Solar Wind Composition, along with the equipment for the field geology experiment, would constitute the rest of the science payload. EASEP would be much lighter than ALSEP and require less time to deploy. By including these experiments on the first mission, NASA hoped to divert the criticism that was sure to come its way and show that its heart was in the right place regarding science. The astronauts would leave the highest priority experiment, the seismometer, at the landing site and still have time to conduct a limited geological study, collecting fewer samples than originally planned.

Instead of being powered by an RTG, the EASEP seismometer would get its power from solar panels and batteries charged by the solar array, the power source rejected for ALSEP but now acceptable because of the short lifetime expected for this substitute. The seismometer would have to operate only through the rest of the lunar day in which it was deployed, although we hoped it might survive longer. It would contain several small isotope heaters to help it survive the lunar night and, with luck, continue to function during a second lunar day. Like ALSEP, it would also have a self-contained telemetry system to transmit to Earth the seismometer and dust detector readings.

EASEP's design was developed through close cooperation between MSC and Bendix, working under the ground rule that as much as possible of the hardware and subsystems would be based on ALSEP. Donald Gerke led the MSC team in the design phase and became the program manager for this hurry-up ALSEP substitute. In November 1968 NASA and Bendix agreed to a $3.7 million contract for the design and manufacture of the EASEP as well as the LRRR. By this time, with the Apollo flight program rapidly moving ahead, the date of the first landing was becoming obvious—sometime in the summer of 1969. Only three more Apollo test flights were scheduled before the first landing attempt. EASEP would have to be built in five months if it was to meet a May 1 delivery date at Kennedy Space Center for a subsequent June or July flight date. In contrast to some of the difficulties we encountered with MSC's ALSEP managers, Gerke was easy to work with, especially for us at NASA headquarters. EASEP proceeded without a hitch and was delivered to KSC one day ahead of schedule.[5]

At the beginning of the chapter I listed the prominent scientists who were identified in Foster's memo, along with the highest priority ALSEP experiments. In the months after the SSSC approved their selection to develop the experiments for Apollo, and before the ALSEP contract was signed, some maneuvering took place—at times a little indelicate—to determine who would be named principal investigators. This title conferred an important imprimatur because the PI would be the primary contact in the years ahead as we built the instrument and also would be responsible for interpreting the returned data and publishing the results. As a reward for all this effort, the PI would receive the largest amount of NASA funding allocated to the experiment and in some cases would be in charge of distributing funds to other members of the team. Remember the golden rule: "He who has the gold rules." This was certainly the case for the PIs. In addition to the gold, they also got the publicity and all the other notoriety that went with this high-profile position. Most prominent scientists are not shrinking violets; being identified as Apollo PIs enhanced their reputations, and the exposure certainly helped advance their careers. How many scientists could look forward to saying they had designed an experiment that was placed on the Moon by the astronauts? Everyone knew only a lucky handful would have this claim to fame.

An example of the competition for this honor was the naming of the PI for the passive seismic experiment. Frank Press, while at Caltech, had developed the first lunar seismometer (which never flew) for Ranger. Maurice Ewing and George Sutton, at Columbia University's Lamont-Doherty Laboratory, had developed a seismometer (which likewise never flew) for Surveyor, and it was this very experience, plus their overall reputations, that led to their inclusion on the passive seismometer team. Ed Davin recalls a meeting at NASA headquarters to select the passive seismometer PI. Press and Ewing were present along with one of Ewing's graduate students, Gary Latham. Ewing, being the senior scientist present, led the discussion and declared that Latham should be the PI because this role would require that someone devote full time to the job and he thought—taking the liberty to speak for himself and Press—that they would not be able to do this. He volunteered that he, Press, and Sutton should remain as coinvestigators. Press, having studied under Ewing at Columbia University, graciously acquiesced, but after the meeting he remarked to Ed, "What Papa Doc wants, Papa Doc gets." He was obviously disappointed at not being named PI by "Papa Doc," a somewhat affectionate name given Ewing by his former students. Soon after, several others would be added to the team, but with Latham at Lamont-Doherty, Ewing's laboratory reaped the public acclaim. Latham went on to do an outstanding job as PI.

I have not had the opportunity to talk to Press about this incident, but I imagine that in hindsight he might think it was not a bad decision. Soon after that meeting he became chairman of the Department of Earth and Planetary Sciences at MIT, certainly a full-time job. His reputation certainly did not suffer from not being an Apollo PI, for among other important jobs he held in later years, he was named president of the National Academy of Sciences, one of the most prestigious scientific positions in the nation.

There are some other interesting anecdotes concerning the experiments. Perhaps the most star-crossed was the Lunar Surface Gravimeter. Its tale of woe has been partially told before, but it deserves further discussion, perhaps with some new insights. I met the PI, Joseph Weber, early in his struggle to get his experiment accepted by NASA. His laboratory was only twenty minutes from our office in downtown Washington, on the campus of the University of Maryland. Dick Allenby, Ed Davin, and I visited Weber in his basement laboratory sometime in early 1964. He had been building and modifying gravimeters in his lab for several years, hoping to arrive at a design sensitive enough to detect gravity waves. Gravity waves were predicted by Einstein's general theory of relativity, and it was believed they could be generated by the collapse of some distant star or perhaps might emanate from an ancient supernova. It was believed that gravity waves would propagate outward from such events at the speed of light and that if one had a sensitive enough gravimeter they could be detected on Earth. Further, it was believed that analyzing them would provide new insights into the structure and evolution of the universe.

A secondary objective of his proposed experiment was to measure the deformation of the Moon by the tidal pull of the Earth. Weber showed us his latest model, and it was indeed a sensitive instrument—so sensitive that it was detecting large trucks and trains passing in the distance. Therein lay the snag that led him to propose his experiment for an Apollo mission. The Earth was subject to so many events that would disturb its gravity field that some thought it would never be possible to make the delicate gravity measurements he wanted. The Moon offered a location without a lot of extraneous gravity sources—certainly no trains and trucks would mask gravity waves. Simultaneous measurements by similar instruments on the Earth and the Moon might identify movements that would be associated only with a passing gravity wave.

Weber's experiment was eventually approved for Apollo, and he was given a contract to build a new gravimeter with the highest sensitivity possible based on the technology of the day (nominal sensitivity one part in 10^{11} of lunar gravity). He in turn contracted with Bendix to build his instrument with a subcontract to LaCoste and Romberg, world-famous builders of gravimeters, to design and supply the sensor. Because of the late approval to include the experiment on *Apollo 17* and the complexity of the design, MSC questioned in August 1970 whether the experiment could achieve delivery in July 1972. We suggested shortcutting some of the normal procedures and, if necessary, integrating the flight hardware with ALSEP at KSC instead of Bendix.[6] Development proceeded on this new schedule with just the usual problems one encountered in such a program, and the flight instrument was delivered on time for integration with the *Apollo 17* ALSEP, the last opportunity to get it to the Moon on an Apollo flight. Because its objective was so unusual, it was billed as the star experiment of the *Apollo 17* mission. Weber and his coinvestigators, John J. Giganti, J. V. Larson, and Jean Paul Richard, all from the University of Maryland, eagerly anticipated being the first to unequivocally detect the elusive gravity waves. Gordon MacDonald, originally on the team with Weber, had dropped out, for reasons I don't recall.

Astronauts Eugene Cernan and Jack Schmitt, aware of its scientific significance, practiced diligently with the training model to be sure they would not foul up its deployment. In his pamphlet *On the Moon with Apollo 17*, printed just before the mission, Gene Simmons, MSC's chief scientist, went so far as to predict that "the practical utilization of *gravitational* waves may lead to benefits that far exceed those gained from the practical utilization of *electromagnetic* waves" (italics in the original). That would be a hard prediction to live up to, but his pronouncement reflected the enthusiasm and anticipation that accompanied the gravimeter to the Moon. An article in *Science* in August 1972 reported that a race was on at a number of laboratories around the world to be the first to confirm the measurement of gravity waves, labeled an "exotic problem."[7]

On the *Apollo 17* mission ALSEP and the gravimeter were deployed on the first EVA by Jack Schmitt, approximately six hundred feet west-northwest of the LM. When commands were sent to the gravimeter to turn on the experiment, readings were received almost immediately back in the Science Support Room. The readings didn't look right to those monitoring ALSEP, but this was the first time the instrument had operated in the reduced gravity of the Moon, so no one was quite sure how the signal should look. Jack completed the ALSEP deployment and activation and went about his other tasks. Meanwhile the Bendix engineers and Weber tried to figure out how to get the instrument to operate more to their liking. They tried to rebalance the beam (the part of the sensor that responded to the pull of gravity) by sending commands to add or subtract mass on the beam, but the signal coming back didn't change significantly with these commands.

A "tiger team" was appointed to come up with a solution while the astronauts were still on the Moon and might be able to help resolve the difficulty, although at the time it still wasn't clear what the problem was or how serious it might be. Perhaps just a little rap by one of the astronauts might clear up what appeared to be a minor discrepancy in the instrument's readings. Schmitt went back to the gravimeter several times during later EVAs to jiggle it a little, but the instrument still did not respond as expected. The beam seemed to be resting on the upper stop and not swinging free. Jack's comments reflected his concern that perhaps he had made some mistake during the deployment, but he had done nothing wrong.

When the *Apollo 17* astronauts left the Moon, Weber and the Bendix engineers were still unhappy with the gravimeter's readings but could not find the cause. Perhaps operating the instrument through one or more lunar day/night cycles might help clear up the signal; so it was monitored for the next several months, but there was no change in the response. The *Preliminary Science Report* for *Apollo 17* came out almost a year later still promising that the gravimeter would return useful information. But it wasn't to be.

Back at Bendix, in Ann Arbor, a second team delved into the mystery. The instrument had been checked out at Bendix before shipment and had worked satisfactorily. What had gone wrong? Then it occurred to LaCoste what had happened. To test the gravimeter on Earth a set of weights were dropped on the balance beam, correctly calculated for Earth's gravity. After the test these weights were recalculated to compensate for the Moon's gravity, which is much less than the Earth's (1/81), and installed by LaCoste. Because of a faulty calculation, those installed were not the proper weights for the Moon.[8] Thus this experiment, on which so much hope for a major discovery had been riding, never returned much useful data. Joe Weber and his team of coinvestigators never forgave LaCoste for the mistake. Perhaps accelerating the schedule contributed to this miscalculation, but at the time it seemed a reasonable risk to get the instrument on the final mission.

At this time, to my knowledge, no one has directly detected gravity waves, and new efforts are under way to build a gravity wave experiment called LIGO.[9] LIGO's announced objective is to detect gravity waves originating from black holes or supernova events. Sound familiar? Designed by scientists at several universities and funded by the National Science Foundation, two identical instruments have been built. One has been installed at the Department of Energy's Hanford, Washington, laboratory and another at Livingston, Louisiana. The two instruments will permit simultaneous measurements at distant points, thus removing the possibility that, rather than signaling the passage of a gravity wave, the mirrors used to bounce a laser beam back and forth in a tunnel two and a half miles long would be misaligned by some local disturbance (such as the trucks and trains observed in Weber's earlier experiments). A difficult quest, but perhaps this time it will succeed.

Another experiment that caused a problem was the lunar surface magnetometer (not to be confused with the Lunar Portable Magnetometer). In this case Chuck Sonett, the PI, chose to have the instrument built by Philco-Ford and then integrated by Bendix. (The PI on the *Apollo 15* and *Apollo 16* missions was Palmer Dyal, also from the NASA Ames Research Center.) The sensor electronics for the instrument contained thirteen hundred active components, eighteen hundred passive components, and thirty-three hundred memory core locations. It included thousands of tiny diodes supplied by Fairchild. Scheduled to fly on the first ALSEP, with the first landing fast approaching, all the ALSEP experiments were under pressure to meet the schedules for delivery, test, and integration at Bendix. Prototype instruments were always tested before building the actual flight hardware to ensure that the design would perform as advertised, and the tests were closely monitored by MSC and headquarters.

When the prototype magnetometer was tested it failed miserably. Short circuits were noted at many places in the circuitry. Trouble. Was there a major flaw in the experiment design? And if so, would there be time to redesign to meet the schedule and have a new instrument ready for this high priority experiment?

The prototype was torn down and subjected to a careful analysis that revealed the problem. To meet the tight schedule, the circuits had not been properly cured, or "burned in," and in addition many of the diodes were contaminated by fine particulate matter embedded in the potting compound. Fixing the curing time was easily solved, but how did the contamination occur?

A team from headquarters and MSC went to Fairchild to review its manufacturing techniques, and the contamination mystery was solved. After the diodes were manufactured, they were placed on shelves—not in a clean room—to cure. Dust and other airborne contaminants circulated in the air and stuck to the potting compound, and these minute particles were enough to permit arcing across the circuits. But could Fairchild come up with a new batch of clean diodes in time to meet the schedule? With the first ALSEP deployment postponed until *Apollo 12*, Fairchild pulled out all the stops and made the delivery, saving the magnetometer's assignment. The instrument operated successfully for many months, with only a few minor discrepancies that were corrected as it continued sending information back to Earth.

Five years after the last Apollo mission, at the end of September 1977, Noel Hinners, who had left Bellcomm and later had been appointed NASA associate administrator, Office of Space Science, decided to save the $1 million per year spent monitoring the five ALSEPs and sending the data to the PIs. Few data were being recorded by this time. It was not expected that the passive seismic experiment, probably the most interesting experiment still operating, would provide much new information because there were no more man-made impacts on the horizon, and naturally occurring major meteor impacts and large moonquakes were uncommon.

During the years they were operating, before being put in a standby condition, all the ALSEPs were still functioning long past their design goals, though occasional glitches and data dropouts were observed. Before NASA terminated support for the ALSEPs, several engineering tests were conducted on the central stations and some of the experiments. These tests were devised to answer questions raised during their operational lives but that had not been allowed to be asked for fear of damaging the ALSEPs and the experiments. The test results were then filed away for possible use if another ALSEP-like station was sent to the Moon. After these tests, commands were sent to the ALSEPs, with each of the PIs sorrowfully taking part in the ceremony, to place their experiments and central stations in a standby mode in case someone wanted to turn them back on later. In the meantime, no data would be collected or transmitted.

In October 1994 the Department of Energy (the successor to the AEC, which had provided the RTGs) wanted to determine if the RTGs had survived over the intervening eighteen years. Ground controllers at the Johnson Space Center tried to reactivate the stations. They hoped the ALSEPs would still be receiving power, as predicted by the plutonium half-life, waiting to spring back to action when Earth called. They made two attempts to turn on each of the ALSEPs, but none of them responded. Although the RTGs were probably still generating electric power, it seems likely that as the RTGs aged and power levels dropped, the ALSEPs turned themselves off, as designed, when a minimum operating power level was reached.[10] The next time they are revisited will probably be when some intrepid lunar explorer or entrepreneur lands near an Apollo landing site and drives over to recover pieces, as we did for *Surveyor 3* during the *Apollo 12* mission, bringing them back to be put in a museum or someone's private collection of space memorabilia.

Walk, Fly, or Drive?

Safety was always the primary concern when someone recommended the astronauts carry out an action. As new ideas were suggested, the astronauts were included as early as possible so they could offer their point of view. When the debates began on how to provide mobility on the lunar surface, they made their thoughts known decisively. The best lunar surface transportation mode would have to take into account not only their preferences but also the payload weight available on the lunar module, the tasks to be performed, and the equipment the vehicle would have to carry. Those looking through the narrow lens of the Field Geology Team wanted the astronauts to cover as much ground as possible at each landing site and carry a variety of tools for mapping and sample collection. The geophysicists and other science disciplines, as we saw at the Falmouth and Santa Cruz conferences, had their own particular requirements for deploying experiments and collecting data. For the Astronaut Safety Office, the primary concern would be to keep the astronauts always within easy reach of the LM in case any of a wide variety of emergencies occurred.

An astronaut walking on the Moon would be, in effect, a small, self-contained spacecraft. His space suit and all the attached systems would have to let him function in the brutal lunar environment (high vacuum, low gravity, and extreme temperatures). It could be as cold as $-260°F$ in shadow, while in full sunlight a short distance away it might be 270°F. He also had to see objects and the ground around him both in shadow and in the glare of the full sun. While moving about he would need a way to maintain voice communication with Earth and, ideally, automatically relay information on his physical condition and the status of his life-support systems so those monitoring them could tell him if he had to return to the LM. Designing a space suit that would accommodate all these multiple functions was an enormous challenge for the Manned Spacecraft Center engineers and their contractors. My office and the scientific community followed their progress with great interest, for the more successfully these challenges were resolved the more scientifically productive the missions would be.[1] The astronauts had to be mobile, and they had to maintain good eye-hand coordination; the closer space suit designs came to allowing "shirtsleeves" efficiency the better, though we knew that could not be achieved.[2]

The space suit solution for the Apollo missions was based on technology developed in the United States and Great Britain, first for pilots flying high-altitude fixed-wing aircraft and, more recently, for the Mercury and Gemini programs. The MSC Engineering and Development Directorate and the Crew Systems Division directed the efforts of many contractors, some retained from Gemini, to produce the Apollo extravehicular mobility unit (EMU), the combination of suit and attached support systems. Hamilton Standard and International Latex Corporation were chosen as the prime contractors for the EMU design and manufacture.

The major elements of the EMU were a liquid-cooled inner garment to remove body heat; an eighteen-layer outer suit, topped by an integrated thermal-meteoroid cover lest a tiny meteorite punch a hole in the suit; a helmet with a clear inner visor and a sunshade (added after *Apollo 14)* and a movable, transparent gold-plated sun reflector visor; gloves; and boots. The portable life-support system (PLSS), attached to the back of the space suit, included batteries, fans, pumps, and the expendables (oxygen, water, and lithium hydroxide canisters to remove carbon dioxide) plus a separate oxygen purge system containing thirty to seventy-five minutes of oxygen in case of a failure in the PLSS.[3] All together, the EMU weighed about 200 pounds (60 for the suit and 140 for the PLSS), varying with the mission and the additions or improvements it embodied. The EMU went through several upgrades from *Apollo 11* to *Apollo 17*, each designed to improve the astronauts' ability to perform their tasks on the lunar surface.

Perhaps most difficult to design were the gloves. I attended several design reviews over the years as improved glove designs, incorporating new materials, were demonstrated. At each review the technology improved, although some ideas were discarded as development proceeded. The gloves had to be tough enough to confine the suit's internal gas pressure (3.7 psi) in the lunar vacuum and to withstand abrasion from handling rocks and equipment. At the same time, the gloves had to allow the astronauts some sense of touch. These two requirements worked against each other from a materials point of view: high wear strength and toughness resulted in poor feel through the gloves. Imagine trying to thread a needle wearing work gloves with the fingers blown up like balloons. Not an exact analogy, but pretty close.

The final design had an outer shell of tough fabric covered with thermal insulation and fingertips made of silicone rubber so the astronauts could feel what they were touching. Not a perfect solution, but the best the technology of the day would permit. In spite of the attention given to this part of the suit, the astronauts would often end their simulations, or return to the LM after a long stint of extravehicular activity on the Moon, with bloody fingertips, cracked fingernails, and their hands aching from trying to grasp and hold a wide variety of objects. However imperfect, the glove design did the job. No glove failures occurred during the missions, and all scheduled tasks were completed.

The EMU restricted how the astronauts could perform various tasks, how far they could wander from the LM, and how long they could stay outside the LM on any EVA. The suit and backpack mass would have to be large, the equivalent of moving a heavy weight with every step. In addition, the astronauts would be continuously working against the internal suit pressure to bend the suit at its joints. Walking on the Moon would thus be difficult and tiring despite the low lunar gravity. If an astronaut fell it was feared he might not be able to get up, and the difficulty was accentuated because the PLSS, attached at shoulder height, raised his center of gravity. (This proved not to be a problem; in the Moon's low gravity, the astronauts could easily bounce up from a fall.) But EVA planning required that they always be close enough together to help each other if one should have a problem. The PLSS provided for sharing oxygen and cooling water if one PLSS malfunctioned.

While suit development was under way, these restrictions raised the specter that the astronauts might not accomplish the demanding work being planned during the lunar EVAs. Metabolic tests had been made on many suited test subjects as well as on several astronauts simulating the tasks to be done on the Moon.[4] Data from these tests showed that the EMU then available would limit EVAs to four hours of low level work. The PLSS could supply consumables (the oxygen, water, and lithium hydroxide mentioned above) for four hours if the astronauts averaged a metabolic rate of 1,200 BTU/hr, the equivalent of playing golf in shirtsleeves. If they exceeded this rate they would have to reduce their activity to reach the average use of consumables if the EVA was to last the full four hours. In reality this would mean almost standing still, since just moving slowly in the suit required over 1,000 BTU/hr; 600 BTU/hr was needed just to work against the suit's internal pressure and overcome joint friction. In spite of improvements in the Apollo EMU during the next few years, the results of these analyses led, in part, to a decision to reduce the amount of EVA time on the first landing mission. EMU consumables were carefully monitored on all missions, especially when the astronauts undertook tasks not programmed in the mission timelines.

These considerations also led to continual upgrades of the Apollo suit and research on better space suits. In May 1968 Sam Phillips asked MSC to recommend a program for space suit development with an eye to improving the astronauts' mobility on the lunar surface for the post-Apollo missions. (He wanted the improved suit to be ready by 1971.) An EVA working group, reporting to Charles W. Mathews, Mueller's deputy associate administrator, began meeting to look into all aspects of EVA, both in free space and on the lunar surface.[5] Ames Research Center became involved, since it also had a team working on space suits; its favorite was the constant volume suit, a hard suit like a deep-sea diver's suit. James Correale led the work at MSC's Crew Systems Division and coordinated the MSC research with that going on at Ames. Many of the concepts combined properties of the soft and hard suits, including articulated bearings, bellows joints, and metal fabrics. Although it promised to reduce the astronauts' workload, the hard suit never was adopted because of operational considerations, including the extra stowage space required. However, the hard suit, or a hybrid suit, is still under consideration for Space Station EVAs because it reduces metabolic demands. Perhaps when materials science improves and spacecraft design permits its use, it will be adopted as the standard EVA suit.

For *Apollo 15, Apollo 16,* and *Apollo 17* several suit improvements were made, including making it easier to bend at the waist and adding expendables (water, oxygen, lithium hydroxide, and a larger battery) to the PLSS to allow longer EVA time—all important improvements for these missions. Since EVAs for these missions might last as long as eight hours, the pressure suits also provided a few creature comforts, with an emphasis on "few." Most important for such long EVAs, bags containing one quart of drinking water were attached to the helmet neck ring inside the suit. The astronaut could reach a straw by turning his head inside the helmet. A small snack bar also could be attached to the neck ring and eaten by turning the head.

At the other end of the human system, a urine bag was attached inside the pressure suit leg to collect urine, much like the earlier "motorman's friend" for trolley car operators. Back in the LM the urine bags would be removed from the suits, and later they would be left on the Moon. Now you know the answer to one of the questions people most often asked the astronauts. The other adjustment made for the final three missions was that some of the tools could be attached to the pressure suit or PLSS so the astronauts did not have to return to the lunar roving vehicle (LRV) to retrieve them from the tool carrier during their sample collecting and geological studies.

EMUs used on the lunar surface EVAs differed from those worn by the command module pilots; beginning with *Apollo 15*, they had to make an EVA to retrieve film and tapes from the experiments bay of the service module during the return trip from the Moon. The CM pilot's EMU did not include the PLSS; it was attached to the CM by an umbilical cord that supplied life-support consumables and voice communication links. The EMU did include a small emergency backpack containing the oxygen purge system, similar to that attached to the lunar surface EMU.

With the Apollo suit being developed, studies described in chapter 3 were already under way at Marshall Space Flight Center on two alternative types of vehicles: flying machines and motorized wheeled vehicles. The wheeled vehicles were championed by most members of the science community, led by the Field Geology Team at Flagstaff, and were supported by my office at NASA headquarters, while the flying machines were favored by some of the staff at MSC and a few astronauts. Our simulations at Flagstaff had used many types of wheeled vehicles, and procedures and operations that took advantage of a vehicle were far advanced. Based on this work, the choice seemed obvious; the astronauts should be equipped with some sort of wheeled vehicle.

Lunar flying vehicle (LFV) proponents at MSC were basing their support on the work that Textron-Bell Aerospace Company had completed at MSFC, also described in chapter 3. The LFV engendered visions of astronauts zooming above the lunar surface like Buck Rogers, free to go wherever they wanted, and quickly. Clearly the LFV would be able to reach places a wheeled vehicle could not go. But would the astronauts be permitted to use such a device, considering safety concerns and the possible need to walk back to the LM from dangerous locations if the LFV failed? Discussions during the Falmouth conference were not supportive of it as an exploration tool. Mission simulations using a flying vehicle were never carried out in the field owing to the difficulty and expense of providing a good simulation. Only Textron-Bell pilots were qualified to use the LFVs, so based on a few demonstrations by the manufacturer, one had to imagine how such a vehicle could be used on the Moon.

This debate came to a head at the Santa Cruz summer conference in August 1967, with heated discussions between the two factions. As is often the case in government matters, when opposing positions are strongly held there are no clear winners, and this was true at Santa Cruz. The final report endorsed both wheeled vehicles and flight concepts. Since we were focusing on post-Apollo missions (in 1967, planning for the first Apollo landing missions envisioned only the astronauts' walking), we were not constrained from advocating robust vehicles, going so far as to recommend using both types to jointly support the surface exploration. In spite of this accommodation at Santa Cruz, momentum was building in favor of a wheeled vehicle for the later Apollo flights. The recommendations coming out of the several working groups called for continuous traverses, manned and unmanned, to sample and deploy various types of equipment and experiments, operations that did not lend themselves to a flying machine.

In April 1969 Frank Press, who had chaired both the Falmouth and Santa Cruz geophysics working groups and was now a member of the Lunar and Planetary Mission Board (LPMB), submitted a paper representing the board's leanings and recommending a "lunar exploration program."[6] Only three months short of the first lunar landing and still anticipating ten lunar landings, Press's paper emphasized the need for enhancing mobility: first, with a better space suit to improve the astronauts' walking and overall EVA capabilities, and second, with some type of wheeled vehicle operating in both manned and unmanned modes to "interpolate between type locations." In Press's words, with increased mobility, the strategy outlined in the paper "provides optimal scientific return and fully exploits the Apollo capability." The LPMB unanimously approved this recommendation at its next meeting in May and passed it on to Homer Newell.

With concerns about the astronauts' ability to move about on the Moon plaguing Office of Manned Space Flight management, George Mueller stepped in and made a decision. The argument of "fliers" versus wheeled vehicles was finally put to rest, and the wheeled vehicle won. Safety was probably the critical factor in the decision. If a lunar "jeep" broke down, the worst result would be a long walk back to the LM. If a flying vehicle had a problem it might crash in an inaccessible area. Other considerations were also important, such as stowage and the overall weight of a fully fueled flier (more than three times as heavy as a projected lunar "jeep") that could carry two astronauts many miles. As envisioned by the Santa Cruz attendees, the LFV would complement a surface vehicle; but as a stand-alone or only means of transportation, the LFV was too limited to support the planned science, especially for the final missions, when multiple EVAs were planned that would include many geophysical measurements at many points along the traverses. Because the LMs had limited payload capacity, a choice had to be made, and the LRV won.

Mueller convened his Senior Management Council in May 1969. At the meeting, attended by George Low, at that time MSC's Apollo spacecraft program manager, and Wernher von Braun as well as other senior OMSF managers, Mueller asked Low and von Braun to examine the problem and arrive at a solution. A small LRV was the final choice, and Mueller told Sam Phillips to go ahead with it. At the end of May Phillips sent a memo to MSFC, the center with the most experience in lunar vehicle research, asking it to manage the procurement. Von Braun wanted an experienced senior manager to lead the effort, and he tapped Saverio "Sonny" Morea to be the program manager. Morea had not been in on any of the earlier MSFC lunar roving vehicle studies, but he had been program manager for the *Saturn V* F-1 engine development, a critical and difficult job that he had successfully completed. He had been given a "heads-up" for his new assignment and had attended the Senior Management Council meeting.[7]

With Morea's appointment, the procurement was put on a fast track. Ben Milwitzky, who had just finished his role as headquarters' manager of the Surveyor program, was transferred to our office to oversee this new program. Ben was a good choice because at the beginning of the Surveyor program a small wheeled vehicle was a candidate payload (though never flown), and Ben had several companies under contract working on their concepts. He had some hands-on experience to guide him in developing the larger vehicle for the Apollo missions.

In July MSFC released the request for proposal (RFP), and three companies responded—Bendix, Grumman, and a Boeing–General Motors team. We all thought Bendix had the inside track to win the contract because of its involvement in all the post-Apollo vehicle studies, plus it was the only one of the three bidders that had a working model of its concept at the time the RFP was released. Boeing also had a good background because of its work in post-Apollo studies, having teamed with General Motors (Delco Electronics Division) for the mobile laboratory competition. Grumman believed it would have an advantage because it had done some earlier work on a one-man vehicle. The design of this new vehicle would be intimately tied to the LM and its stowage constraints, and of course no one knew the LM better than Grumman.

After the Source Selection Board (SSB) reviewed the proposals, it determined that Bendix and Boeing had the superior proposals and passed its findings to NASA headquarters. Because of the short schedule—seventeen months from projected contract start to delivery of the flight vehicle—headquarters told MSFC to negotiate contracts with both companies, not knowing which one would be chosen by the source selection official, Thomas O. Paine, the new NASA administrator. With negotiated contracts in hand, we would be able to jump-start the contract and save valuable time. Of the two bids, Boeing had submitted the lower price, $19.7 million, and since all the other SSB findings were essentially equal, Paine awarded the contract to the Boeing team.

MSFC then signed a performance-based contract (a wise decision, as it turned out) that went into effect in November 1969. Included on the Boeing-GM team were Eagle-Pitcher Industries, which supplied the LRV batteries, and United Shoe Machinery Corporation, which provided the electric harmonic drive units that powered each individual wheel. It would be a true four-wheel-drive vehicle. The contract called for the delivery of four vehicles (later reduced to three) and six test units, one of which was eventually converted into a one-g trainer for astronaut simulations on Earth.

Soon after the contract went into effect, MSFC and headquarters had some misgivings about the specifications contained in the contract. Morea's team thought they were too complex and opened the door for possible change orders that would boost the price and perhaps jeopardize the schedule. For example, the original RFP called for a gyroscopically controlled navigation system. After careful review, the high accuracy this type of system would deliver was thought to be unnecessary, and it would add to the overall cost. On January 15, 1970, Ben chaired a meeting of engineers from MSFC, MSC, and Kennedy Space Center to rectify this situation and develop a less restrictive set of specifications.

The design requirements coming out of that meeting, and then translated into the final specifications for the Boeing team, called for an LRV that would carry one or two astronauts plus experiments, communications, a TV camera, and crew equipment and would provide stowage for lunar samples collected during the traverses—a total payload capacity of 970 pounds.[8] In place of the gyroscopic navigation system, it would have a rudimentary system that would give the astronauts a continuous vector back to the LM in case it was out of sight and they needed to make a rapid return. Other specifications called for the LRV to travel a maximum of ten miles an hour on level mare surfaces with an overall range of seventy-two miles.

The most demanding requirements were that the vehicle be transported to the Moon in the wedge-shaped LM descent stage Quadrant I and that the total weight of the vehicle, including its stowage and deployment mechanisms, could not exceed four hundred pounds. This meant the LRV would have to be folded or collapsed and that the chassis and wheels would be flimsy indeed.

After all the vehicle studies we had performed for the post-Apollo missions, I was skeptical that the overall specifications could be met within the weight and stowage constraints. This would be smaller and lighter than anything we had studied for post-Apollo, yet it was being designed to accomplish many of the jobs we had envisioned for our larger vehicles. I shared my concerns with Ben, but he was convinced the specifications were valid. Events proved that such a vehicle could be built with these tight constraints. I credit his management skills, along with the dedication and engineering know-how of Sonny Morea's team plus the hard work and cooperation of Boeing, GM, and their suppliers, for the on-time delivery of the LRVs—the payload stars of the last three Apollo missions.

The LRV team encountered many complications as it struggled to meet the tight schedule. Early in the contract, MSFC concluded that the Boeing program manager did not have the skills to manage such a critical program and asked that he be replaced. Boeing agreed and brought in a new manager, Edward House, who took control and saw the project through to its successful conclusion. The next problem was the escalating cost. Congress got wind of this and asked the Government Accounting Office to review the contract. Here the performance-based contract proved valuable, because MSFC could demonstrate that the contractor's rising costs were justified, based on the LRV's design complexity, and that the contractor fee (profits) would be adjusted accordingly to arrive at the best price for the government. At a hearing at which Milwitzky and Rocco Petrone, who had recently replaced Sam Phillips as Apollo program director, testified, they explained the way the contract worked. They were able to satisfy the House Oversight Committee that the costs were realistic for such an unusual vehicle. The matter was dropped, and the final cost, with modifications to the original contract for the LRV flight and test units, was just under $37 million—a bargain in the opinion of all who were involved in the missions.

While the LRV was in development, two new data points were thrust into the discussions on astronaut mobility. The first was the comments of the *Apollo 11* astronauts after their return. Although their EVAs had been reduced in number and length so that their total time on the surface was just a little over two hours and thirty minutes, Neil Armstrong and Buzz Aldrin came back with the impression that walking on the Moon would be easy. They had discovered that a loping, rolling gait was the most efficient way to move and helped overcome some of the space suits' deficiencies—in particular the difficulty of bending at the joints. Armstrong said he thought an LRV would not be needed to get around and to conduct the tasks the scientists had planned. When Morea asked at one of the debriefings what size wheels he would recommend to ensure that the LRV could handle surface irregularities, Armstrong replied, "about twenty feet."[9] His opinions carried some weight, but in the end they did not slow the development of the LRV, and a much smaller wheel (sixty-four inches), did the job.

The second, more positive data point was the experience of the *Apollo 14* astronauts. For *Apollo 14* we had built a small two-wheeled cart called the modularized equipment transporter (MET) that the astronauts would pull along loaded with whatever equipment they needed during their traverses and that would also store the collected samples. By this time the array of geological tools and sampling devices we wanted the astronauts to carry had grown considerably, including three cameras. As Alan Shepard and Edgar Mitchell struggled to reach the rim of Cone Crater, the primary sampling objective of the mission, the MET became a bigger and bigger hindrance. In the end, as they tried to climb the slope to the crater rim pulling the MET behind them, they decided it was easier to carry it. Walking and pulling even a small cart created such a high workload that the astronauts often had to stop and rest before continuing their exploration. Because of the extra effort expended attempting to reach the rim, and with time running out, they were forced to return to the

LM, and they never quite reached their objective, though they came close. There seemed to be no question that with the much more ambitious missions next on the schedule, we were right to insist on having a motorized vehicle to carry the astronauts and their equipment.

By the time the first LRV was delivered to KSC on March 15, 1971, two weeks ahead of schedule, some of the original specifications had changed. Overall weight had been allowed to grow to 460 pounds, and its allowable payload had also grown, to 1,080 pounds. Its total range had decreased from seventy-two miles to forty. The reduction in range was acceptable as new mission rules developed for the LRV traverses dictated that the astronauts stay within six miles of the LM so they could walk back if the LRV failed.

Television pictures and voice communication would be possible from the LRV at the limits of the traverses, out of sight of the LM. A self-contained lunar communications relay unit would be carried on the LRV or could be hand carried. The LCRU would provide a direct link to Houston by two antennas mounted on the front of the LRV. The low gain antenna would permit voice relay with only coarse pointing toward Earth, but the high gain antenna, required for TV transmission, had to be pointed rather accurately by the astronauts. This meant that voice communication would probably be available throughout an EVA, but TV pictures normally could be transmitted only when the LRV was stopped or when driving if the antenna happened to be pointing toward Earth. The LCRU would also permit a operator at Mission Control to point and focus the TV camera when the astronauts were working away from the LRV. The first LRV would be available starting with *Apollo 15,* and we were waiting with great anticipation for the TV pictures from the new LCRU. It promised the flexibility to monitor and communicate with the astronauts that we had tested in our post-Apollo simulations at Flagstaff.

Edward Fendell, who got the nickname "Captain Video," trained for many hours to operate the TV camera from his station in the Mission Operations Control Room during our Apollo simulations and had become adept at manipulating it to get the best coverage. This skill was invaluable to the "backroom" Field Geology Team, and Ed cooperated to the fullest with their requests for views of the local topography at each stop. The media, especially the TV networks, were also excited about closely observing the astronauts at work and broadcasting live the promised spectacular scenery of the last three landing sites.

As a bonus, the LCRU would let us witness an LM takeoff from the Moon. At the end of the last EVA, the astronauts would drive the LRV about three hundred feet from the LM and park it with the LCRU on board and the TV camera pointed toward the LM. If Fendell could coordinate elevating the camera with the liftoff, we would be able to watch the LM disappear into the black lunar sky. Despite the difficulty of slewing the camera fast enough to follow the rapidly accelerating LM, Fendell accomplished this feat. At the end of the *Apollo 15* mission, the world saw for the first time a slightly blurry view of a spacecraft taking off from another body in our solar system. We were also able to see the effects the LM's ascent engine exhaust plume had on the lunar surface and the Apollo Lunar Surface Experiments Package. It was a little frightening for the ALSEP engineers to see debris flying in all directions, but the ALSEP survived. If the LCRU still had enough battery power after the *Apollo 15* astronauts left, we hoped to take pictures of the lunar eclipse that would occur a week later (assuming the launch stayed on schedule, which it did), as well as other views of the lunar surface and astronomical targets. These observations were successfully carried out.

A few final words will describe the LRVs, the remarkable machines that made *Apollo 15, Apollo 16,* and *Apollo 17* so successful. The wheels were constructed of an open wire mesh, to reduce weight, make it easy to stow in the small LM bay (the wire mesh was compressible), and damp the ride by flexing and acting as shock absorbers as the LRV bounced across the lunar surface in the low gravity. The open mesh had some drawbacks, however; as was correctly predicted, the wheels picked up soil and sprayed it over the LRV and the astronauts, so each wheel was covered by a small fender to direct the spray downward. (On *Apollo 17* one of the fenders came loose during the first EVA traverse, and the soil spray coated the LRV and the astronauts' space suits and equipment with a thick layer of dust. The next day Gene Cernan and Jack Schmitt made a new fender by taping together stiff sheets from their landing site maps and attached them over the wheel. Even so, when riding on the LRV or just walking around, the astronauts would return covered with lunar soil that they had to brush off before reentering the LM.

The LRV's front and back wheels could be steered together, in tandem, or each pair independently, allowing it to make tight turns. It was steered with a small T-shaped hand-grip controller, which also regulated speed and braking. A knob below the T-handle controlled forward and reverse, much as in a golf cart.

Mounted above and just forward of the T-handle was the control and display panel, which contained a speedometer, LRV system switches (e.g., for power and steering), temperature gauges, and the onboard navigation system. This last system provided a continuous bearing and range back to the LM and also showed the total distance traveled to help the astronauts find their predetermined science stops.

All in all, the LRV was a dandy little machine that performed flawlessly. Full-scale models can be seen at several NASA centers as well as at the Smithsonian Air and Space Museum, which also displays a lunar module mock-up and other examples of equipment the astronauts used. If—or when—we go back to the Moon, it would surprise me if small vehicles similar in appearance and performance to the Apollo LRV are not part of the equipment included in the payloads. Why pay to redesign such a successful system? I hope Boeing or NASA has kept the drawings.

Astronaut Training and Mission Simulation

Just before I arrived at NASA, in April 1963 the United States Geological Survey had reached an agreement with the Manned Spacecraft Center to start a geological training program for the astronauts. Ellington Air Force Base, a few miles west of the proposed location for the main MSC campus and home of the NASA astronaut air force, was selected as the site for this rump USGS office. Gene Shoemaker chose Dale Jackson, a former marine, to lead this effort, thinking his background would allow him to mesh successfully with the astronauts, who were all military pilots. Until that time the astronauts were not perceived as enthusiastic about studying geology, in view of their other pressing duties. By the time I joined NASA, stories were already circulating that some MSC staff members and Jackson's small team did not agree on who was to call the shots on this important function. MSC staffers believed they should be in charge, although USGS had been given this mandate by NASA headquarters. Adding to the problem, the newly hired MSC staffers assigned to work with Jackson's people did not have as much experience as Jackson's staff, yet he agreed to include them in the training. As in other areas I have described, MSC had a pronounced fear of being left out of important assignments related to Apollo science and tried whenever possible to monopolize these roles.

In spite of the friction between the two staffs, Jackson plowed ahead with his duties and devised classroom and fieldwork courses in basic geologic principles, mineralogy, and petrology. With the astronaut office's approval, the syllabus called for fifty-eight hours of classroom lectures and four field trips. The fifty-eight hours of "geology" training were part of an overall classroom syllabus of 239 hours designed to prepare the astronauts for the upcoming Gemini flights.[1] The geology training was not related to the upcoming Gemini flights,

159

the astronauts' primary concern at that time, and would not have real value unless they were selected as Apollo crewmen. Thus it was not universally embraced, especially by some of the original seven and the second and third astronaut classes. Eventually, however, it became accepted as an essential box to be checked off if one hoped to be chosen for a Moon mission. It was anticipated that after crews were selected for the lunar landing missions, five additional series of follow-on lectures and field trips would be scheduled.

By 1967, one hundred hours of classroom lectures and ten field trips became the requirement for astronaut geology training. This training, and then the mission simulations, would become more and more rigorous and realistic as the program matured and simulations were scheduled using prototype and final design equipment and tools.

Three weeks after joining NASA in September 1963, I attended my first demonstration of a prototype Apollo space suit at MSC. The demonstration and briefing were done under the auspices of MSC's Crew Systems Division. Hamilton Standard had been awarded the overall contract to develop the Apollo space suit and backpack, with International Latex, its subcontractor, responsible for the suit itself. This was my first opportunity to see the current state of the art in space suits. The prototype Apollo suit we were to see demonstrated was the latest amalgamation of this technology, plus modifications added by the Crew Systems staff, which had the ability (or expertise) to second-guess the contractor and make its own adaptations when appropriate. At this point two types of suits were under consideration: a "soft suit" made of multiple layers of nylon and other material and a "hard suit" to be made of some type of hard plastic or honeycombed aluminum material. This was a "soft suit" demonstration, the preferred approach.

A test engineer wearing the suit went through a series of mobility exercises for the assembled throng. Some movements he could carry out easily; others were more difficult or almost impossible. Bob Fudali and Noel Hinners of Bellcomm also attended the demonstration and filed a detailed report on what they had observed. They wrote: "All in all, it looks as if mobility will be rather low (even in improved suits) and that the astronauts will not travel far from the LEM without additional mechanical aids. [Their] ability to set up equipment and perform experiments on the surface will also be quite limited unless striking changes are made in future suits."[2] I also reported in a memo to my office what I had seen and what I believed were the deficiencies in the design.

My first exposure to astronaut training and simulation came at the end of August 1964 with a trip to Bend, Oregon. At this early date many had questioned the astronauts' ability to carry out meaningful scientific observations and work on the lunar surface while encumbered by the available space suits. I was one of the skeptics, based on the earlier space suit demonstration at MSC. My report on the 1963 demonstration had gotten back to Max Faget's office at MSC and was considered so negative that when MSC found out I would be attending the Bend simulation, Faget sent a telegram to Tom Evans disinviting me. Ed Andrews told me to ignore the telegram and go anyway.

The Bend simulation, supported by several MSC offices, was designed around a space-suited astronaut, Walt Cunningham, alternating with two MSC technicians in space suits. They would work at several locations, using a few rudimentary field tools, and at the same time report what they were doing and seeing. The Bend location was chosen because it seemed like a good terrestrial analogue of what the astronauts would find on the Moon. It consisted of three types of volcanic terrain. One site was primarily a field of basaltic extrusives, jagged and rough and in places containing pieces of obsidian. MSC, it was rumored, was considering using the area as a permanent simulation site. Governor Mark O. Hatfield (not yet a senator) and the press had been invited to witness parts of the simulation, and the exercise rapidly turned into a major public relations gaffe.

During the simulations, Walt wore the prototype Apollo space suit demonstrated less than a year earlier, with a few improvements including a new backpack. It was the best suit available at the time. Together the suit and backpack and a bulky white overgarment weighed more than a hundred pounds. It was a blazing hot day, uncomfortable even for those of us just standing and watching in shirtsleeves. Walt's suit was fitted out with a new water-cooled inner garment, best described as a pair of long johns with a network of thin plastic tubes sewn on. Cold water circulating through the tubes was supposed to keep him from overheating. It didn't. His visor often fogged over, and he had trouble seeing where he was going.

One slope he tried to climb was covered with pieces of razor-sharp obsidian, and as might be expected, he tripped and sliced a hole in one of his gloves. Before this he had tried to use a geologic hammer and scoop to pick up samples. Both tasks were awkward in such a garment, but to make matters worse he had to carry the tools in one hand or hung at his waist and at the same time manipulate either a "walker" or a "Jacobs staff" that was supposed to help him conquer this rough terrain. At every stop he would put down the walker or staff and begin his next task. No matter how hard he tried, every action looked difficult. Whenever he bent over he tended to lose his balance because the suit was not designed to bend easily at the waist, a deficiency we had noted a year earlier. After he fell and cut his glove he continued to tumble down the slope and was saved from injury only by two technicians standing nearby just in case. All in all, it was a simulation disaster, which the local press reported the next day in large headlines.

By the end of the simulation, with a short rest after his fall while the tear in his glove was repaired ("duck tape" helped get us to the Moon), Walt attributed his problems to his fogged-over visor and other suit limitations. He described the scene to his superiors back in Houston as a "Roman holiday," referring to the swarming photographers eagerly taking pictures of his pratfall. Bob Fudali of Bellcomm also was there to observe the simulation. In his report he noted that "predicting the mobility of an astronaut on the lunar surface from these tests would be a serious error."[3] My report to my office also retold Cunningham's mishaps, and when copies of our memos were brought to his attention, he came to associate us with his bad press. The main points of our memos had been to argue for a suit that would make the astronauts more mobile and for better-designed tools, not to criticize Walt's efforts. This simulation was an important factor that led him to caution us at the Falmouth summer conference not to overload the astronauts with lunar surface science tasks. Later I was able to explain my position to him and we became good working partners, though Walt never quite forgot his embarrassing Oregon experience.

My report also addressed the disadvantage of having such a large public attendance at simulations where many new things would be tried for the first time. I recommended that future simulations be done at Flagstaff, where we were beginning to set up good facilities and where attendance might be controlled. I had, of course, an additional motivation: to legitimize the role USGS was playing in our post-Apollo simulations and put the staff in a position to more strongly influence what would be done for Apollo. Will Foster and E. Z. Gray agreed with my suggestion, and each sent a memo to George Mueller recommending that Flagstaff be the future site for simulations.[4] The Office of Space Medicine also sided with our observations and recommended policies to guide future simulations, including that astronauts "not be used as test subjects" unless they would make some unique contribution.[5] Mueller forwarded these memos to MSC. He got back a letter from George Low, deputy director at MSC, disagreeing with Foster and Gray on their recommendation to conduct future field simulations requiring special terrain at Flagstaff and claiming there was no intent to set up a "lunar training camp" at Bend.[6] This last statement played down Governor Hatfield's comments while he was at the simulation that he supported having such a "camp" at Bend. This seemed to confirm the rumors we had heard that MSC had indeed made some preliminary overtures. It was clear that Low was telling Mueller they intended to do their own thing, especially when dealing with USGS.

Low's response prompted Foster to send Mueller another memo to clear the air; he said that his earlier memo was not intended as a criticism of MSC but repeated his concern that pressure was being exerted on NASA to establish a training facility at Bend.[7] To put an end to this internal bickering, Mueller wrote to Bob Gilruth, the MSC center director, "It is my desire that the Centers work closely with the USGS . . . and that there be no unnecessary duplication of field simulation activities," and he sent an identical letter to Wernher von Braun at Marshall Space Flight Center.[8] This exchange, unfortunately, only deepened the growing animosity between MSC and our headquarters-USGS team.

As field geology training picked up speed and our post-Apollo studies progressed, we were constantly trying to find sites that would demonstrate terrains similar to those we expected the astronauts to encounter on the Moon. USGS already had a selection of sites it used at different stages in the training program, depending on the objective. Training trips took the astronauts to many distant places, both in the United States and overseas. But as our understanding of the Moon grew from pictures returned by Ranger, Surveyor, and Lunar Orbiter, new sites that could mimic the lunar surface were in demand for both Apollo and post-Apollo mission planning.

In May 1964 Bill Henderson, Don Elston, William Fischer of USGS, and I went hunting for sites that might be suitable for simulating longer missions and lunar base activities. Final reports from Bill Henderson's Lunar Exploration Systems for Apollo (LESA) lunar base studies were due in nine months. Interim reports were already suggesting a broad range of undertakings that could be carried out at a base, and we used these early reports as a starting point for planning lunar base simulations. In those heady days we were thinking big; a lunar base program would undoubtedly be announced in the near future, to follow the successful Apollo missions. Until this time simulations for post-Apollo missions had been conducted exclusively near Flagstaff. We were looking for one or more large sites, not too remote and preferably on government property, where we could expect to find support for the lunar base simulations, which we anticipated would be complex. We drew up a list of potential locations, obtained photographs and other background material, and reduced the large number of candidates to a short list.

We went first to the Atomic Energy Commission's Nevada Test Site (NTS), where a series of surface and subsurface atomic and high energy chemical explosive tests had pockmarked the landscape with craters of all sizes. The local AEC manager was interested in our proposal, and though the site had restricted access, some sections could be made available for training. We were given a helicopter overflight, and from the air there was no question that it appeared moonlike. One crater, called Sedan, was especially impressive. Formed by a 104 kiloton explosive, the crater was 320 feet deep and 1,280 feet across. Flying over it at low altitude reminded me of standing on the rim of Meteor Crater in Arizona, for it had many of the same characteristics. After we landed we toured the site by truck to get a closer view. When we got out of the truck at the first stop, we discovered a major problem; we had to put on white coveralls and boots because the surface soil was still slightly radioactive; the atomic clocks of some of the products of the nuclear explosions were still ticking. We should have expected this situation, but when we made our calls to set up the tour, the fact was not mentioned. We looked at each other and rolled our eyes, then after a few short excursions we thanked our hosts politely and left.

Our second stop was China Lake, a large navy test range in southern California. We studied a large-scale map of the range at the headquarters building and selected a few spots for a close-up truck survey. The range was vast (1.1 million acres), with lots of room for the many exercises we were hoping to conduct. Although it was not as Moonlike as NTS, vegetation was sparse and there were many interesting geological formations that could simulate lunar conditions. We toured the range by truck and agreed that it looked like a good site, and the commanding officer seemed willing to accommodate us. The test range also included many shops, hangars, and other facilities that we would need to support long-staytime simulations. They could be made available, we were told, with appropriate compensation.

From China Lake we next visited Fort Huachuca, Arizona. After a meeting with the commanding general, who assured us of his interest, the army also provided a helicopter overflight, followed by a series of briefings on facilities and other advantages of working there. They were definitely selling: perhaps they saw reduced budgets in their future and thought this new use might offset these reductions. This army proving ground was beyond question isolated. The Huachuca Mountains formed the western border of the fort, and a variety of volcanic terrains could be found within its boundaries. Although the region was semiarid, it was a "green desert." Most of the ground was covered with cactus, including cholla, palo verde, and other types of plant life common to the area; it was beautiful, but we thought it would be too difficult to cope with continuously for sustained long-distance walking and vehicular simulations.

Our final stop was the White Sands Missile Range in south-central New Mexico. It was similar in many respects to China Lake. There was lots of space, some areas had Moonlike terrain, and there were good support facilities. NASA was already using some of the range, so we would not be unwelcome guests. It was perhaps the best of the sites we visited. As events unfolded, we never had to make a choice. Lunar base funding and planning came to an end about a year later, and our more modest post-Apollo simulations were all carried out near Flagstaff.

We continued to look for additional Apollo training sites, however, and a new tool became available to assist us. On each Gemini flight the astronauts took photographs of the Earth's surface with handheld Hasselblad cameras. Many showed areas never before well documented with aerial photographs. For each flight Paul Lowman, with his coinvestigator Herbert Tiedemann at MSC, had designated points of special interest that the crew should try to photograph, time permitting. Gemini missions were launched due east from Kennedy Space Center to take full advantage of the extra boost from the Earth's rotation; thus their flight paths repeatedly covered all of the Earth's surface from 28.5° north latitude to 28.5° south. One of the benefits of repeating the launch inclination was that it was possible to rephotograph the designated areas when the photos from earlier missions were of poor quality or were not taken. This also allowed some stereoscopic coverage where the photos overlapped.

Using these photos, Paul and I searched for other potential training sites. Each Gemini photo typically covered an area of some 3,500 square miles, with the oblique photos covering even more—an unprecedented continuous view of the Earth's surface. In the typical aerial survey, an average frame might cover less than ten square miles. Conventional photographic coverage of the large areas included in a typical Gemini frame would require constructing photo mosaics, with trained photogrammetrists piecing together many separate photographs. Having used such products in our geological pasts, we knew that no matter how skillfully fabricated, photo mosaics always introduced false information in the finished maps. A geologist could be misled by something that looked like a stream or valley or some geological feature such as a fault but was really an edge between two photos.

Features never fully photographed before the Gemini missions, such as the Richat structure in Mauritania, that might be the result of large meteorite impacts were of special interest because they might provide not only training sites but also the opportunity to learn more about impact processes. In 1965 only a few well-documented impact craters were known throughout the world, and many of them were so obscured by erosion that they were not well suited as training sites. Thus we were constantly trying to find more examples that we could study or use to train the astronauts.

A few of the Gemini photos had been published in *National Geographic, Life,* newspapers, and other publications, but the vast majority had not been seen by the general public. In his spare time Paul had been carefully cataloging the pictures and interpreting their geologic content. It occurred to us that these new views of the Earth might interest companies exploring remote parts of the world. So far, no commercial interest had been shown. If we could get a positive response, it would support NASA's proposed Earth orbital remote sensing program—just in an early planning stage—and perhaps persuade NASA management to accelerate this program.

In May 1966 I called Mobil Oil in New York and talked to my old boss, James Roberts, who had been transferred after I left Colombia, first to Venezuela and then to Mobil headquarters. I explained what we had and what we thought would be the potential benefits and applications of space photography. He said he was interested in seeing the photographs and agreed to set up a meeting with some of the Mobil Exploration staff, the unit responsible for finding new oil fields. A few weeks later Paul and I flew to New York to show the Gemini photos to their first commercial audience. We brought to the briefing some of the best examples of geological features photographed by the astronauts; mountain ranges in the southern Sahara (Mobil was heavily involved in exploring remote areas in Libya and Tunisia) and clear pictures of structures in Iran of the type petroleum geologists looked for (anticlines and synclines). I knew Mobil had several field parties working in Iran at that time, because before I left Colombia Iran was a possible new destination for me. We also included a few spectacular views of the Andes and the Himalayas. We felt sure there were no aerial photographs of some of these areas, and this would be the first time Mobil had such views available. We thought they would be impressed.

We were wrong. For whatever reasons, the staff members Roberts brought to our meeting showed little interest. They said they had, or could get, enough conventional coverage so that space photographs were not needed. This response mystified us. Perhaps they thought an endorsement would leave them open to providing financial support for an undertaking with an uncertain future. We will never know what might have happened if Mobil had been enthusiastic. Like other programs that were struggling to get started at this time, the Earth orbital observation program limped along, in part because there was no strong commercial interest. It would be many years before the unmanned Landsat program and Skylab would be launched.

Our search for terrestrial impact structures took us on two trips, one back to Colombia in April 1964 and another to Peru in June 1968. We visited Colombia to study a small circular structure of unknown origin, Lake Guatavita, high in the eastern cordillera of the Andes, some thirty miles north of Bogotá. Lake Guatavita was an intriguing and well-known feature; at the time of the Spanish conquest it was rumored that the Chibcha Indians, who lived on the high plateau that surrounds what is now Bogotá, used the lake for special ceremonies. It was said that the local chief would cover himself in gold dust every year and then bathe in the water, accompanied by other sacrificial ceremonies. The Spanish had dredged the lake and attempted to drain it in hopes of finding sunken treasure. A modern attempt, again unsuccessful, had also been made to drain the lake after several marvelously intricate gold artifacts were recovered from the bottom. Geological study had failed to come up with a satisfactory explanation of the lake's almost perfectly circular shape; one suggestion was that it was created by an impact, but no proof had been reported. I had visited the lake while living in Colombia and was aware of its history and the impact theory.

Now that there was better understanding of how to identify an impact crater in the field, Paul and I developed a field study plan for making a quick assessment of the lake and submitted it for approval. The estimated cost of the trip for the two of us, including all expenses, was $1,000. In the memos that went back and forth before approval was given, a number of interesting comments were appended to the routing slips. The most humorous was one made by George Mueller's special assistant, Paul Cotton: "George, this is the slickest justification for a boondoggle I have ever seen. As long as we have this kind of resourcefulness, we should be confident of reaching the moon and planets." A second staff comment to Mueller was that approval should be given only if we included an astronaut. We were in favor of this recommendation, but it was soon shot down as taking too much valuable astronaut time. Our "resourcefulness" was rewarded, and the trip was approved.

Our plan was to quickly survey the lake's immediate surroundings looking for evidence of impact in the form of shatter cones or other impact debris such as ejecta, glass, or meteorite material. For two days we tramped around the half-mile-diameter lake picking up samples, taking pictures, and making a few measurements. We could find no evidence of an impact. This left us in a quandary: How should we report our results when there was so little to report? We felt sure that thin-section study of our samples would only confirm our field observations that the lake was not the result of impact. We went back to my old Mobil office in Bogotá to examine more closely what was known, geologically, of the immediate area. Based on the published literature, we concluded that since we could find no evidence of an impact the lake was probably formed when the surface rock collapsed over a small salt dome that had been dissolved by groundwater. Thick salt deposits were known to exist in the underlying formations, and a complete cathedral had been carved below ground from the salt at Zipaquirá, a short distance away. And so we reported our findings.[9]

When E. Z. Gray forwarded our report to Mueller we received a short handwritten acknowledgment: "I doubt if the returns were worth the time and money. Do you agree?" Gray wrote back: "What value do you place on developing an organization? I am a firm believer in learning by doing. I think this trip was worthwhile." Although it was only a small incident in a rapidly accelerating major national undertaking, this story provides a measure of the attention to detail demonstrated by senior management and at the same time the freedom of action they allowed their staffs. Such management competence, and such security in their abilities, may have had no equal in a government program before or after and was, I believe, instrumental in Apollo's success.

The Peru trip was instigated by our study of the photographs returned by

Gemini 9. During the flight the astronauts had photographed the Andes from Chile to Colombia. At the point where the mountain chain turns from a mostly north-south direction to the northwest near Lake Titicaca in southern Peru, we observed several large circular structures, each having a diameter of thirty miles or more. Were they created by impacts or by some other mechanism?

After Paul and I found the circular structures on the Gemini photographs, we tried to determine if they had been discussed in the geological literature. We found no citations. Such large structures, if formed by impacts, would be a major discovery. We could see many large impact craters on the Moon, and by this time we had in hand the detailed Lunar Orbiter photographs that showed some of the fine structure associated with large impacts. We knew of no impact craters of this size on Earth, although we were sure that, like those on the Moon, they had been made during the planet's early history. The Ries Kessel structure in Germany, about fifteen miles in diameter, which was used as an astronaut training site, was the largest confirmed terrestrial impact feature known at that time. The Vredefort Dome in South Africa, some twenty-five miles across, was potentially a larger example but was yet to be studied in detail. Many aspects of the large lunar craters were intriguing, especially their central peaks. Only large lunar craters had such peaks. Why did they exist? Did they reflect the thickness of the lunar crust or some other unknown phenomenon? The Gemini photos showed that the large circular structures in Peru had mountains in their centers. We started to lay plans to visit Peru and try to answer our questions on the origin of these features.

As our planning progressed, Paul could see it would be difficult for him to make the trip; he had returned to Goddard Space Flight Center and new duties. I continued to pursue the idea and finally received permission to go from my new boss, Lee Scherer. In preparation I had been in contact with the United States and Peruvian embassies as well as the Peruvian Geological Survey and was assured of their cooperation. From the Defense Intelligence Agency I had obtained aerial photographs of the area taken in 1955 so I could plot our findings in the field. Interestingly, these relatively high resolution individual photographs gave no indication of the structures, and a photomosaic made from these photos also failed to show them. The advantage of the small-scale space photos, which covered a large area without distortion, was clear. In addition to these rather formal arrangements, I received an unexpected bonus. A NASA colleague, Rollin Gillespie, who worked in the Planetary Missions

Office, was interested in joining me. His son Alan, who was majoring in geology at Stanford, was also interested; so Rollin, at his own expense, offered to meet me in Lima and accompany me along with several Stanford students.

I arrived in Lima on June 15 sans baggage and field equipment, lost somewhere en route. Rollin and his group had arrived several days before and had been in touch with the Peruvian Geological Survey. He had already made arrangements for two Land Rovers and for drivers, guides, translators (Spanish to Quechua), and three Peruvian geologists to accompany us. This saved us several days, since I arrived on the weekend and could not have made such connections for two days. While waiting for my baggage we met with the minerals attaché at the United States embassy and with several other organizations that were conducting mining operations in the area, and they supplied important information about the conditions we would encounter. An engineer at the Madrigal Mining Company told us they were working several large copper and silver mines in the center and on the flanks of two of the structures. This was encouraging; perhaps these circular features were similar to the Sudbury structure in Canada, thought by some to be the remains of an impact crater, which was being mined for nickel, copper, and other metals.

Our plan was that Rollin and I would fly to Cuzco, where we would be joined two days later by the rest of the party and the Land Rovers and then travel south to the site. We flew to Cuzco on schedule and met, as we had arranged, with geologists at the National University of San Antonio to explain our project. They had never seen the Gemini photos and were excited by them. They were familiar with the region but had never realized these circular structures existed. While visiting at the university we received our first bad news. The rest of the party had been delayed in leaving Lima and would not arrive for several days. We decided to have them bypass Cuzco and meet us at Sicuani, a town near the base of the mountains. Before leaving the university I promised to stop on my way back to Lima and lecture to faculty and students on the Apollo program.

The next day Rollin and I took a bus to Sicuani, the only "gringos" on a bus filled to capacity with local passengers and all their baggage, some of it alive. It was essentially a straight shot through the Vilcanota Valley, which connects Cuzco to the altiplano that surrounds Lake Titicaca. Sicuani lay some eighty-five miles south of Cuzco by way of unpaved roads but with some spectacular scenery along the way. We arrived in Sicuani late in the afternoon and checked into the only hotel (warm water available every morning from 7:00 to 7:30). It was very cold. Sicuani is at an elevation of 12,000 feet, and there was no heat in the rooms, where we spent an uncomfortable night. By chance, while walking in the main plaza that first night, we met an American Carmelite priest who invited us to the parish house, where we discussed our plans with the assembled fathers. We then received our second round of bad news. They had visited the general area and told us it was not possible to drive in—it was too rough and there were no roads. We would have to rent horses. This would certainly slow up our exploration and add more time than I had available. They suggested we enlist the bishop's support.

We met Bishop Hayes the next morning, and he was very helpful. Not only did he understand local politics and know who could ease the way, but he had a large, comfortable house (hot water all day) where he invited us to stay. We immediately agreed. The rest of our party arrived the next day, and we completed our arrangements for renting horses and obtaining other equipment. With the delays in getting started my time in Peru was running out. I would be unable to travel to the structures and would have to depend on Rollin and the Stanford students, along with the Peruvian geologists, to complete the survey.

Returning to Cuzco by train, I stopped for the afternoon to deliver a lecture at the university. From Cuzco I flew back to Lima and then home. Back at NASA, I received a package from Professor Carlos Kalafatovich V. on the staff at the university in Cuzco. It contained several Peruvian newspaper clippings noting that scientists from NASA had visited the region and were interested in the mountains near Sicuani. According to the papers, which featured big black headlines that translated to "Flying Saucers Land in Canchis" (a small town near Sicuani), some of the local people interviewed were intimately familiar with those mountains. It seems that the locals knew of frequent visits by flying saucers that came to extract precious gems from somewhere in the mountains and take them back to their home planet. Now we knew what had attracted us to these structures.

On a more serious note, the party I left behind was not very successful. It was almost impossible to travel in the mountains, even using horses. They collected a few samples and took them back to Stanford for analysis. They found nothing unusual, and no sign of impact was observed in the mineralogy of the returned samples. The origin of the circular structures was not solved, and as far as I know the question is still open.

Backing up a bit, in September 1965 I participated in one of the astronaut training trips to Medicine Lake, California, a site near several small, complex volcanic features. By this time astronaut training trips were well organized by USGS and included prominent geologists who could lecture and teach the astronauts about the importance and subtleties of the locations selected and about their potential similarities to lunar features. This was the second two-day trip astronauts made to the area, and those on this particular trip were Russell "Rusty" Schweickart and Roger Chaffee. Roger was soon to be named to the crew selected to fly *Apollo 1,* scheduled to be the first manned flight of a Saturn rocket. Gene Cernan was also scheduled for this trip, but because of a hurricane threat he was delayed in Houston and unable to attend.

Roger Chaffee had come to the astronaut corps from the navy and held the rank of lieutenant commander. Since we were both jet pilots with many similar interests and experiences and had flown off some of the same class aircraft carriers, we hit it off immediately, and he became my truck mate for the training trip. I drove, and between scheduled stops and lectures I would fill him in on geological lore I thought he should know. But as I remember, we mostly swapped sea stories about night carrier landings and the idiosyncrasies of the planes we flew. He seemed to welcome the change of pace from his "normal" astronaut assignments, even though each day he was subjected to nonstop lectures and fieldwork while being force-fed textbook geology.

The team assembled for this trip consisted of ten people. Aaron Waters led the team and was to deliver the lectures and coordinate the trip itinerary. He was supported by nine helpers, including three USGS camp hands, two USGS geologists, two MSC geologists, and two MSC photographers. The astronauts' doings were always well documented by photographs. Dick Allenby and I were also invited for this trip, so there were fourteen of us. We slept in one- or two-man tents and were up at dawn to complete each day's tightly scheduled business. Breakfast was served around a campfire because the early morning hours were already chilly. At noon we had box lunches, and dinner was back at the campsite. This trip turned out to be especially memorable because William Rust, one of the USGS "camp hands" but in reality a technician, was the designated cook and an inveterate fisherman. Each morning, before any of us were awake, Bill would go to the lake and catch trout, then cook them for breakfast—a treat in any circumstances but for these few days a Washington bureaucrat's delight.

Roger Chaffee's attendance was especially significant and attested to the astronauts' growing awareness of the importance of these trips as well as to Roger's personal interest. Usually astronauts who would soon receive flight assignments could not take time off to attend to business other than that directly related to their flights, and there definitely was no geology to be done on *Apollo 1*. Roger enjoyed the training and was becoming an able field geologist. I'm sure he hoped word of his new skills would get back to Deke Slayton and Al Shepard and put him in line for future Moon missions.

I told him I intended to submit my application for the next scientist-astronaut selection and hoped I would soon join him in the astronaut corps. Neither Roger's flight nor my selection came to pass; less than two years later Roger died tragically in the *Apollo 1* fire along with his two crewmates Virgil "Gus" Grissom and Edward White. Their deaths led directly to a major re-evaluation of how NASA was preparing for the Apollo missions, however, and the changes in the way NASA would do business ultimately ensured the program's success.

Here is as good a place as any to relate my own experience in attempting to become an astronaut and give some idea of how scientist-astronauts were selected. Although I had been a military pilot, as were almost all the astronauts, I didn't have a lot of jet hours; most of my flight time had been logged on propeller aircraft many years earlier. After working with the astronauts for a year and knowing their flight backgrounds, I could see that it would be virtually impossible for me to qualify in a typical selection process because I lacked current piloting experience. Then I heard that scientist-astronauts might be recruited. In April 1964 NASA asked the National Academy of Sciences to develop procedures for selecting them. Gene Shoemaker had lobbied for such a selection, and before he was diagnosed with Addison's disease he had been considered a probable top choice when NASA finally got around to agreeing it needed such positions. Even after knowing he would not be selected, Gene continued to lobby, and his efforts, along with those of others in the science community, eventually paid off. I bided my time feeling that my best chance to qualify for the astronaut corps would be through the scientist-astronaut program.

When the call for applications was finally announced in October 1964, I quickly obtained the packet with the paperwork to be completed. It listed standards for such qualifications as age, height, and educational background.

Height! Maximum allowed height was six feet. I was six feet one. The age limit excluded anyone born before August 1, 1930. I was nine months overage. I made a few calls to see if these requirements were inflexible and found that they were. The height restriction was based on the dimensions of the Gemini capsules and the Apollo equipment then under design, which would not comfortably accommodate anyone over six feet. Greatly disappointed, I wrote to the National Academy of Sciences, the initial screening hurdle, to tell them I was interested but was disqualified because of my age and height, and that I hoped these restrictions might one day be changed so that I and others in my predicament could apply.

The good news about this first scientist-astronaut selection was that Jack Schmitt, then working on projects we were sponsoring at Flagstaff, made it all the way through, and he and five others became the first of this special group. Suddenly we were to have a strong advocate in Houston, someone who saw eye to eye with our concerns; but we would have to wait a year for his help while he trained to be a pilot.

I had written to the Academy with deliberate forethought. I felt sure there would be other scientist-astronaut selections. Our post-Apollo planning at that time called for extensive scientific experiments on the lunar surface, and qualified scientists would have to perform them to satisfy the scientific community. George Mueller had testified before Congress on these plans, and I knew he supported the need for additional scientist-astronauts. My letter, I hoped, would be retrieved at the next selection, showing my long-term interest in the program and perhaps influencing the selection criteria.

To give myself a better chance in the next selection, whenever it might be, I decided to apply for a pilot slot in one of the Navy Ready Reserve squadrons at nearby Andrews Air Force Base. My last flying experience had been with a navy reserve squadron in Denver while attending graduate school. No pilot openings were available at Andrews in 1964, so I joined an intelligence unit drilling once a month to get back in the Ready Reserve flow and learn through the grapevine where pilot assignments might be found.

This contact soon turned up a vacancy at Lakehurst Naval Air Station, and I quickly transferred to VS-751, an antisubmarine squadron, to resume flying after a seven-year layoff. A year and a half later, with new flying time under my belt, I persuaded a fighter squadron commander at Andrews who needed pilots to have me transferred, and I began the transition to the F8U Crusader. But the navy got wind of this behind the scenes activity; needing antisubmarine-qualified pilots, it rescinded my transfer and assigned me to VS-661 at Andrews. Although I was disappointed (I was looking forward to flying the Crusader, one of the navy's best-ever fighters), the transfer had one redeeming factor. I would now fly out of Andrews and save the long monthly commute to Lakehurst. And at least I was flying and could hope that this would be a plus in the next selection.

In September 1966 the National Academy of Sciences announced the second scientist-astronaut selection. Accompanying the press release was a short statement by Gene Shoemaker, who would be chairman of the Academy's selection panel: "Scientific investigations from manned space platforms and direct observations on the Moon will initiate a new phase in man's quest for knowledge. While such missions call for daring and courage of a rare kind, for the scientist they will also represent a unique adventure of the mind, requiring maturity and judgment of a high order." Who could resist such a challenge? I thought that, with Gene as chairman and knowing several other members of his panel, I would have a real chance. It was rumored that this would be a larger class than the previous group of six, thus improving my odds. The Academy had been somewhat disappointed by the number of applications received for the first selection, although the six chosen had excellent qualifications, and thus the selection criteria were a little more relaxed the second time. The age and height limitations had not been changed, but this time the press release stated that "exceptions to any of the . . . requirements will be allowed in outstanding cases." Perhaps now I had a chance. Could I qualify as an "outstanding case"?

My application must have been one of the first received. As I remember, almost five thousand applications were screened for this second selection. Evidently there had been enough good publicity about the Apollo program in the interim to encourage many young scientists to want to be a part of it. About two hundred were selected for the next phase of physical and psychological examinations; I made the cut. We were divided into small groups and sent to the Air Force School of Aerospace Medicine at Brooks Air Force Base in San Antonio, where all astronaut candidates were screened.

We endured a week of prodding, blood work, and spinning, IQ, and many other tests, some of which were vividly shown in the movie *The Right Stuff,* though not with the same comic detail. (For a more complete account of what we experienced, read Mike Collins's book *Carrying The Fire.*) While I was tilted upside down with my stomach filled with a barium solution, they discovered that I had a slight hiatal hernia; the muscles in my esophagus couldn't hold all of the solution in my stomach. Because it was apparently a minor ailment and because, I assume, the other test results were good, I was sent to the Walter Reed Medical Center in Washington, D.C., for a second opinion. The examination at Walter Reed went well, and the examining doctor wrote a letter to NASA saying he did not consider the diagnosis disqualifying—that at the worst I might have to take an antacid to relieve any discomfort I might feel in zero gravity.

Where did this leave me? I couldn't be sure, but I did have enough experience to know that astronaut selections were secretive. I knew Deke Slayton and Al Shepard were involved, but I didn't know who else. By this time I was acquainted with all the astronauts, including Al and Deke, but I wasn't sure whether this was good or bad. I had been on field trips with them, from time to time I was invited to brief the astronauts on the plans for post-Apollo missions, and I was often in the astronaut office building to visit Jack Schmitt and other astronauts as well as the Crew Systems staff. I felt I had a good relationship with them, but perhaps my differences with some MSC managers might hamper my selection. In June I received the call I had been hoping for. I had made the final cut and was invited to Houston for the last interviews before a selection was made.

In June 1967 twenty-one candidates made this final visit. A few of them I knew from my week in San Antonio. Their backgrounds included almost all scientific disciplines, but as I read the list I saw I was the lone geologist, along with one geophysicist. Only two earth scientists! Most of the post-Apollo science activities we were planning had some earth science connection; I thought my selection was in the bag. The first scheduled activity after checking in was a ride in a T-38, the astronauts' aircraft of choice, based at Ellington Air Force Base. This was a piece of cake. I flew the plane from the front seat with a NASA pilot (perhaps evaluator?) in the back seat. I did some simple maneuvers and a few snap rolls and generally showed off my flying skills. From what I read in the brief bios of the other candidates, I believed I was the only one with experience as a jet pilot. If this was a test, I must have passed. Next we took a ride in the MSC centrifuge; as I remember, they spun us up to about six gs while we performed a few simple exercises of hitting some light switches. Not a problem, and I suspect some of our future bosses were looking on through closed-circuit television to see how we did on the nearest thing to a stressful test.

After a few other briefings came *the* interview. I recall only four people in the room: Al, Deke, Bill Hess, and Charles Berry, who was head of the medical sciences office—"the astronauts' doctor." All the questions were rather innocuous. Berry asked about the hiatal hernia, and since I had seen the Walter Reed report I told him that I hadn't even known I had it until the test and that I didn't think it would cause any trouble. The only question that stands out in my mind was the one Deke asked: "Don't you think you're too old to be an astronaut?" I was thirty-seven at the time and not the oldest of the final twenty-one candidates, but I knew I was over the advertised age allowance, so I had done a little homework. I answered, "I don't think so; after all, I'm younger than Wally Schirra, and he's still flying." This brought a big laugh from all four inquisitors. Considering that Walter Schirra, then forty-three, was the only astronaut from the original seven to fly in all three programs—Mercury, Gemini, and Apollo— my answer was evidently on the mark. That ended the interview, and Al said he would give me a call. I thought my selection was now only a formality. That afternoon I did some preliminary house hunting in the neighborhoods around NASA.

In August Al called. "Don," he said, "I'm sorry to tell you you weren't selected." We talked for a few more minutes, and I'm sure he realized my disappointment. They had chosen eleven for the scientist-astronaut class of 1967, including the geophysicist Anthony England, the only other earth scientist. I didn't ask why I wasn't selected; I was sure he wouldn't give me any specifics. I rationalized that it was a combination of things. My hiatal hernia (they didn't have to take any chances on its causing a problem); my seniority (from a government classification standpoint I would have been senior to most of the astronauts selected earlier); my pilot background, which may have been seen as a negative (I would have been the only one they didn't have to send to pilot training, and that might have made me an apple among all the oranges. What would they do with me during the year the others were in training?) Finally, they might have received some negative comments from MSC managers I had disagreed with in years past.

Alan Shepard died recently, so I won't get a chance to ask him why I wasn't chosen. Perhaps he would have told me, perhaps not; most probably, after so many years he wouldn't even have remembered. In any case, the rejection probably did those of us not selected a favor from a career standpoint. Within three years the post-Apollo missions, the prime reason for the selection, were canceled, and none of the class of 1967 flew on a space mission for fifteen years; Joseph Allen was the first from this class to fly as a mission specialist, on shuttle flight STS-4. A few retired or left NASA before taking part in any NASA missions, and several, like Joe and Story Musgrave, made major contributions to NASA programs.

Returning to training and simulations, geological field training for the astronauts became more and more realistic and intensive as the date for the first landing came closer. By 1966 all the astronauts had had some level of both classroom and field training. Those in the first three groups selected had the most extensive geological training. Since no one knew who would ultimately be selected for the landing missions, we tried to have them all at as high a level of competence as possible within the time available. Many noted geologists volunteered to assist in the training; some stayed on to become members of the Apollo Field Geology Team and worked with the astronauts until the last mission, *Apollo 17*, was safely home. Lee Silver, Richard Jahns, Aaron Waters, Dallas Peck, and William Muehlberger come immediately to mind as volunteers who devoted a significant part of their professional careers to these efforts. Many others made important contributions to astronaut training, including many geologists on the staff at MSC.

I was able to take part in several field geology training trips, and those I attended were all memorable. A specially arranged visit to the Pinacate volcanic fields in Sonora, Mexico, just over the border from Arizona, had a somewhat different purpose. This trip took place in late summer 1966. The Pinacate area includes an interesting set of volcanic craters formed by the explosive release of superheated underground water; craters of this type have their own geologic name—maars. From the air they have an uncanny resemblance to some lunar craters: their rims are only slightly raised, the craters themselves are symmetrical, and many are relatively shallow. Some of those at the Pinacate are quite small, a few hundred feet across, and two are very large, the largest being over one mile in diameter. The area where they occur is desolate and isolated, a perfect place to take a high profile group like astronauts, where no one would disturb their training. (It was definitely a place where reporters would not go, for there were no amenities of any kind.)

The Pinacate became one of the favorite training sites, and most of the astronauts made a visit at one time or another. This visit was without astronauts; its purpose was to educate my bosses, Phil Culbertson, who had replaced Tom Evans in August 1965, and his boss, E. Z. Gray. Since we were still looking for new training sites for the post-Apollo missions, I thought it was important to show E. Z. and Phil how we would use such sites and what benefits could derive from good terrestrial analogues like the Pinacate. I had arranged with Gene Shoemaker and Al Chidester to conduct the trip as if it were an astronaut training trip, with Phil and E. Z. being treated, in a manner of speaking, as the training subjects.

For both of them it would be a real eye opener; we would camp out in tents for two days in the middle of nowhere, something they had seldom experienced. We all flew in to Phoenix and were met by the USGS staffers who would support the trip. Then in a caravan of four or five trucks we turned south on Route 85 with a first stop at Ajo. At that time Ajo was a copper company town with a company store that sold provisions at a discount; the USGS guys always knew how to save a buck. Among other food, we bought frozen T-bone steaks to grill over an open fire the first night; with no refrigeration, we had to cook them that day, and by the time we made camp we expected they would be thawed. From Ajo south, Route 85 takes you through Organ Pipe Cactus National Monument, a unique desert habitat with numerous large saguaro cacti standing like statues along the highway and stretching off into the distance in all directions. This was the "green desert," with all kinds of unusual plant life including mesquite, palo verde, cholla, and other thorny stands of wicked-looking cactus that I had first seen when we visited Fort Huachuca.

We crossed the border at Lukeville and turned west on Mexico Highway 2. Almost immediately the landscape changed dramatically, becoming much more barren and arid with only a few scattered houses along the road out of Sonoita, the small Mexican town opposite Lukeville. After a few miles we turned off on a dirt road and continued south; the dirt road turned into two tire tracks, and finally we drove in and out of the dry arroyos, gaining a little elevation, and arrived at the volcanic fields about three in the afternoon. While the USGS support team set up camp, we walked over to the rim of Elegante Crater for our first look at the next day's simulated training site. Elegante Crater is impressive. Over five thousand feet in diameter and eight hundred feet deep, it was not unlike Meteor Crater in many respects, except there were no large blocks of ejecta around the rim and few blocks or large boulders in the interior. The crater looked as though it had been scooped out of the desert by a large spoon, and whatever had been in the center had disappeared. These craters normally constituted a difficult test for the astronauts to interpret and describe so that the accompanying geology staff, acting out the role of a support team back on Earth, could develop a reasonable geologic map based on the astronauts' descriptions.

By the time we returned to camp the tents were all set up and a campfire was lit. Gordon Swann and I went back to the pickup for the frozen steaks and lifted the cardboard carton to carry them over to the cook. They had thawed, the carton had turned to mush, and the thirty or so steaks fell through the bottom onto the sandy soil. What a mess. With a carefully rationed supply of drinking water to last the two days, we could spare only a little to wash off the steaks, so they were still crusted with sand when they finally hit the grill. E. Z. and Phil, along with the rest of the crew, were treated to a new dinner sensation: steak that wore your teeth down if you bothered to chew. I could tell E. Z. wasn't enjoying his outing—not the best way to impress the bosses with how well organized we were on astronaut training trips. Around the campfire that night the veterans of this type of trip told tales of previous visits to the Pinacate and described some of the exploits they had been party to. Some of the astronauts were enthusiastic card players, and apparently a few exciting card games on past visits had gone on into the wee hours, affecting their next day's concentration and ability to absorb some rather detailed geological lectures. As we knew, not all the astronauts took the field training seriously.

We were a much more sedate group than some in the past, except that a couple of USGS staffers had brought the makings for powerful after-dinner drinks. By the time the storytelling was in full swing, several in the cast were oblivious to the heat and sand. Those of us who were not imbibing heavily decided to call it a day, and along with Phil and E. Z. we crawled into our tents. With fewer seniors around the campfire to dampen the storytelling, the talk grew louder and louder, punctuated from time to time by the equivalent of an Arizona rebel yell. Finally E. Z. couldn't take the noise any longer. He jumped out of his tent and threatened to cut off all USGS support if they didn't immediately shut up and go to bed. This got their attention; the noise decreased to a low rumble and then silence. When we finally fell asleep, all we could hear was the buzzing of the night insects.

The next morning, up with the sun, we were gathered around the fire awaiting breakfast and the first geology lecture when we noticed that two staff members were missing. We searched around the campsite and couldn't find the midnight revelers. We were getting worried; rattlesnakes, scorpions, and gray wolves inhabited this area, and there was even an occasional panther. Finally we found one of them asleep in a truck cab, and the other turned up several hundred feet from the camp, lying near a clump of cholla, slightly the worse for wear with his shirt torn and a little bloody. Thus was added another chapter of tall tales for future astronaut training trips. But for E. Z. Gray it was the last straw; he cut his visit short and was taken back to Phoenix that afternoon. By the time I got back to Washington he had calmed down, and we continued to support USGS in all its work. Training trips to the Pinacate were considered highly successful, and on missions to the Moon some of the astronauts would comment on how much the Moon's surface looked like their memory of the Pinacate.

Mission simulations for crews assigned to specific Apollo lunar landing flights had a somewhat different aspect. For these exercises the two astronauts assigned to the lunar module would be involved, often with their backup crew and sometimes with the command and service module crew member, depending on the objective of the simulation. This meant a support crew of dozens. In addition to the astronauts, lecturers, and technicians, the ever present MSC photographers would be milling around snapping pictures from all angles. Walt Cunningham's simulation at Bend, Oregon, was an intimate gathering (with the exception of the press that was present) compared with these later simulations. As we approached the flight date, simulations would progress from casual dress at analogue field sites to full suited simulations at MSC or KSC, with some of the latter attempting to follow projected lunar timelines as closely as possible.

As principal investigators were identified for each of the science experiments, they would also attend from time to time, along with the contractors building the equipment, so they could observe how the astronauts deployed or operated their instruments. At times the simulations would result in changes to accommodate the astronauts' ideas on how to improve their interaction with the particular experiment; but whenever possible the astronauts attempted to adjust to the idiosyncrasies of the experiment and achieve the best results for the PIs.

By this point in the training (crews being selected for specific missions), the simulation sites included an MSC high-bay building, the "back lot" at MSC, a small outdoor site at KSC, and a few special analogue sites scattered around the country, chosen to be most like what the astronauts would find on the Moon. The MSC "back lot" or "rock pile" was a few acres of simulated lunar terrain with an LM mockup in the center. The surface was covered with gravel and sand and salted with various types of rocks. A smaller simulated outdoor lunar surface was built at KSC, primarily as a convenience for the astronauts, who spent more and more time there as their launch date approached. The KSC site was often unusable because the "craters" would fill with water at high tide (very unmoonlike), but this site permitted last-minute reviews of specific tasks that may have been added or modified since the previous simulations at MSC. The KSC outdoor site did not include an LM mock-up, so it could support only limited types of simulations. However, there was an indoor site that did include an LM simulator. The KSC simulations were usually conducted in pressure suits to be as authentic as possible. Equipment provided was spare flight article hardware or the closest copy we could obtain.

One of the special analogue sites was near Sunset Crater, a few miles northeast of Flagstaff. Calling it an analogue is a bit of a misnomer, because it was in fact the closest copy of a moonscape that existed anywhere on Earth. Some of the staff at Flagstaff hit on the idea of duplicating the lunar surface as seen in one of Lunar Orbiter's pictures. They carefully analyzed the selected frame, measuring the diameter and depth of all the small craters and interpreting the history of this small piece of the lunar surface by determining the relative age of each crater based on how the ejecta layers overlay each other. After these calculations were made, Norman "Red" Bailey and Hans Ackerman, two Astrogeology staffers, laid out a grid of fertilizer bags on a ten-acre volcanic ash fall south of Sunset Crater. When the fertilizer and fuel oil explosive was detonated, the Orbiter photo was recreated. Not only were the bags arranged according to the explosive force they would generate to create the proper size craters in the correct locations, but they also were timed to go off in the sequence that would provide the correct ejecta layers observed on the real lunar surface. It was a roaring success in all respects, and the creation day was delayed until I was able to witness it on one of my frequent trips to Flagstaff. A movie was made of the explosions, and it was great fun to replay it for visitors who came to watch the astronauts training at the site; each new crater erupted in sequence, in slow motion, and the fine ash flew skyward in great dark jets.

This site, and two additional sites formed in the same manner, became the last tests for the astronauts, requiring them to use all the observational skills they had gained. As they walked or drove around on the closest thing to the Moon they would see until they actually landed there, they described it to the backroom crew so that a geologic map could be made. After completing the exercise, they would review their observations with their instructors to correct any misinterpretations they might have made. All the astronauts from *Apollo 13* onward trained at these sites, and I always thought it was one of the best simulations they were involved in, since it was the most complete test of their skills at observation and description.

A drawback with all the pressure suit simulations was that we could not replicate the one-sixth gravity field they would experience on the Moon. In some sessions we tried to simulate the low lunar gravity by using two types of simulators and specially rigged harnesses that partially suspended the test subject and reduced his weight to one-sixth of his Earth weight. These simulations were usually not very satisfactory because the complicated harness setup would reduce only the astronaut's apparent weight, not the weight of the equipment he was working with. But some of these tests provided important insights, since the mass of the equipment was accurate and the astronauts got a feel for this unique combination of forces. The NASA airplane, normally used to simulate low or zero gravity, also was a poor substitute because of the short duration of each flight parabola. Neutral buoyancy simulations (held in a tank the size of a swimming pool)—a much better way to simulate low gravity environments and the standard way to train for today's shuttle missions—were in their infancy. They were used for simulating the zero gravity parts of the missions, but not for lunar surface tasks.

In addition to simulating the geologic tasks they would carry out, the astronauts simulated the deployment of the Apollo Lunar Surface Experiments Package and the use of all the other equipment and experiments they would carry on the mission. For the final three missions the important equipment additions were the lunar roving vehicle and the lunar drill. The LRV's deployment from its stowed position on the LM landing stage became a critical part of the timeline. To accomplish all the tasks planned for the extended-staytime missions, the astronauts had to get the LRV functioning as quickly as possible. This meant removing it from the LM stowage bay and setting it on the surface while simultaneously unfolding the wheels tucked beneath the frame, erecting the TV and communication antennas, and finally checking the drive system to be sure it had survived the long journey. A clever but complicated system of cables, springs, and hinges was designed for the LM and LRV.

Once they were sure the LRV was operating correctly, they would load it with the other equipment and experiments that depended on the LRV for their operation. LRV deployment was rehearsed over and over again to reduce the time it took and try to ensure success. During the training sessions the MSC and KSC staffs would introduce hang-ups in the deployment of the LRV and other equipment to see if the astronauts could overcome such adversity. They soon became adept at doing this and foreseeing problems.

Another important task to simulate was getting the loaded lunar sample return containers back into the LM from the lunar surface. This maneuver tested the ingenuity of the MSC engineers because the astronauts could not carry the bulky containers up the LM ladder. They devised a pulley system. One astronaut would kneel in the LM hatch while the other stayed on the surface to hitch the containers to the pulley cables and slowly pull them up to the waiting astronaut. Although it was a relatively straightforward solution, the cable system tangled easily, so it took many hours of practice to rig the pulleys and coordinate the two astronauts' actions. Lending urgency to these "rock box" simulations was the knowledge that of all their tasks this was the most important—the harvest of Moon rocks and soil. If for some reason the sample containers were left behind, the mission would be deemed a failure. This would be especially true for the final missions, which would include samples from locations far from the lunar equator and precious cores collected from below the lunar surface by the lunar drill.

By mid-1967, detailed training and simulation schedules were set up for each of the lunar landing missions.[10] Starting forty-four weeks before their scheduled launch date, the astronauts would follow a tight schedule designed to cover all aspects of the missions. Almost 2,200 hours of training and briefings were crammed into their workdays at both MSC and KSC. Some required the presence of all three astronauts, others called for the CSM pilot alone, or just the two Moon-landing astronauts. This constituted a scheduled fifty-hour workweek for each of the three astronauts and the backup crew, with untold extra hours of unscheduled time. They underwent a minimum of 5 hours a week of physical training, 6 hours a week of flying time, 5 hours a week of Apollo flight plan reviews, and 25 hours of flight-suit fit checks, 196 hours of spacecraft tests, 20 hours reviewing stowage procedures for both the CSM and the LM, 40 hours of planetarium exercises to ensure that the crew could use celestial navigation to update their programed navigation system in case of several possible failures, 10 hours of egress training to cover water recovery from the CSM after splashdown, 269 hours of briefings and simulations for science operations, and many other types of training. The 269 hours of science training was one of the largest time allocations, and it was jealously guarded by those of us involved in providing the science payloads, since the other side of the NASA house—the engineers, flight controllers, and other critical participants in launch preparations—would try to preempt some of this time for their own use. But in spite of this constant demand for more astronaut time to attend to nonscience matters, Deke and Al stuck to the schedule, and we were seldom shortchanged. After being named commander for *Apollo 14,* and while involved firsthand in the training for his mission, Al became a strong supporter for the science team's training requirements for the final three missions.

When the contract was signed to build the LRV for the last three missions, Rutledge "Putty" Mills, our vehicle guru at Flagstaff, was charged with building a training vehicle that would approximate the LRV configuration so that we could continue to do mission planning and simulations at Flagstaff. (The flight version of the LRV could not be used in terrestrial simulations because it was designed to operate in lunar gravity. It would have collapsed under the astronauts' Earth weight.) An LRV simulator that could be used in Earth's gravity was not due from the contractor for some months, and we wanted to get an early start on our simulations. Putty did his usual innovative job of constructing a vehicle from odds and ends and his fertile imagination. We named it "Grover the Rover," for one-g rover, and it was ready for testing by the end of June 1970, just six months after Boeing was given the final LRV specifications. At the end of August we conducted a full-scale test, with astronauts participating as well as others. Astronauts in attendance were John Young, Charles Duke, Tony England, Gerald "Jerry" Carr, William Pogue, and Fred Haise.

The test was scheduled to be conducted at the Cinder Lake Crater Field Number 1, but most of the driving over the next four days took place at a vacant lot near the USGS building in Flagstaff. The astronauts present operated the Grover, as did engineers from MSC, MSFC, and NASA headquarters. Putty had built the Grover to run on electric motors like the real LRV, and he had three battery packs available to recharge so we could have more or less continuous operation. At full throttle the Grover could make seven miles an hour carrying two passengers, similar to what we could expect of the LRV on the lunar surface. Mock-ups of some of the tools were stowed on a pallet on the vehicle, the way we anticipated they would be carried on the Moon, although a final stowage configuration for the LRV had not been decided. At the end of the test, all agreed that the Grover would be a valuable addition to future mission simulations, especially when Putty had a chance to add refinements such as a navigation system and additional mock-ups for the lunar communications relay unit, TV, and other equipment the LRV was scheduled to carry.[11] Eventually we obtained a fully functional spare LCRU for our simulations.

A site selected for the simulations conducted toward the end of crew training for the final missions was on the island of Hawaii. Despite the prevailing view that most lunar features were the result of impact processes, all the astronauts had visited Hawaii early in their geologic training to study the wealth of lunar-like features created by the many active or semiactive volcanoes. Simulations for specific missions were a different matter, more like a final exam. We chose several locations on the island to represent geological situations similar to those the crew might encounter on the Moon. Typifying the Hawaiian simulations, the *Apollo 17* crew spent the first four days visiting these sites, then had a day of rest. Dallas Peck, a noted volcanologist who had spent a number of years in Hawaii studying the island's geology, acted as coordinator and principal lecturer. The final three days were spent at Kahuku, Hualalai, and the volcanic ash wastelands at the crest of Mauna Kea (elevation 13,796 feet), chosen to represent what astronauts Gene Cernan and Jack Schmitt might find at their designated lunar landing site, the Taurus-Littrow Valley.

At Mauna Kea the staff had prepared a series of traverses around the volcano's summit that would approximate those the crew would follow on the lunar surface. Sampling and description stations had been designated at intervals replicating as closely as possible the Taurus-Littrow timeline that had already been carefully plotted by the Field Geology Team for the actual mission. All the surface equipment the crew would deploy or operate, except for ALSEP, was transported to the top of the crater, including a simulated version of the LRV. Putty Mills had modified a local jeep to use as a simulated LRV, a cheaper and less sophisticated version of the Grover and other LRV training vehicles. It also avoided the expense of transporting one of these trainers from the mainland to Hawaii. He had removed most of the jeep's body and engine so that the astronauts were sitting on open seats on the frame and could climb on and off easily. He had also added racks for their tools and sample bags and a mount for their communication antenna, similar to the stowage on the real LRV.

During this training exercise most of us lived in motels on the coast, either in Hilo or in Kailua-Kona, commuting the thirty to forty-five miles a day to the training sites. Some of the USGS staffers lived closer in an army base and kept most of the equipment we would use each day there. Cernan and Schmitt wore street clothes for these simulations; it would have been too costly and time consuming to try to conduct them in pressure suits this far from Houston. To add some mission reality they wore backpacks similar to the portable life-support system, but with battery power only for voice communication back to our simulated Science Support Room out of sight of the traverses.

Bill Muehlberger, the Field Geology Team PI appointed for *Apollo 16* and *Apollo 17,* was in charge of this trip. He brought several members of his team including George Ulrich, Gerry Schaber, and Dale Jackson. Scientist-astronaut Robert Parker was also on hand, since he had been designated mission scientist and the prime capsule communicator during the periods of extravehicular activity. Muehlberger and his team would man the rudimentary SSR, connected to the astronauts only by radio, plotting their progress as they drove around the summit and communicating through Parker as they would during the actual mission. The Field Geology Team, through trial and error on earlier missions, had devised procedures to assist the astronauts if something unexpected happened or to respond to any questions they might have, and these procedures were also practiced.

Those of us not directly involved in the backroom simulation would follow Cernan and Schmitt from a distance as they drove from station to station, making note of how everything fit together—or didn't, as the case might be. At the end of the exercise, Muehlberger and his team retraced the traverses with Cernan and Schmitt, reviewing how they interpreted their voice reports, correcting their map, and then suggesting ways to improve the crew's descriptions to produce a better interpretation of what they actually saw.

With the first scientist-astronaut geologist in the crew and a highly motivated and well-trained commander, we didn't expect there would be much need for this type of support, but as with all things NASA, we were going to be prepared. All in all, this Hawaii simulation was about as good as we could get in obtaining a high fidelity rehearsal before the real mission was under way.

We conducted one week of intensive, almost uninterrupted training for both the crew and the Field Geology Team. *Apollo 17* would be the last mission, and Muehlberger was determined that it would be the best if he had anything to do with the training and simulations. In just five months it would be the real thing. A final reward for our efforts had become a tradition. On the last night of these trips, a dinner was held at Teshima's, a lovely Japanese restaurant high on a hill overlooking the ocean, with Mrs. Teshima providing a royal welcome and a special menu. It was a night of storytelling, practical jokes, and reminiscing, a dinner that all who attended will long remember.

10

Studying the Moon from Orbit

Although the Ranger and Surveyor missions had sent back many close-up views of the lunar surface, they were never intended to provide all the photographs we would need to select the Apollo landing sites. That was to be the job of Lunar Orbiter. Conceived in 1963, its objective was to obtain detailed photographs of the whole Apollo landing zone. We needed high resolution in order to pick areas free of large boulders or small craters that would be a hazard to the astronauts guiding the lunar module to a safe landing. Obstructions of this size could not be seen on photographs taken from Earth, even by the largest telescopes. The Lunar Orbiter program was managed by the Office of Space Science (later the Office of Space Science and Applications), but the photographic design requirements were dictated by the Office of Manned Space Flight and in particular the engineers at the Manned Spacecraft Center. Langley Research Center (LaRC) was selected to be the day-to-day manager, and the request for proposal was released by LaRC. The RFP called for building six to eight orbiters; it was possible that the final ones in the series would include other experiments in addition to cameras. OSSA released an announcement of flight opportunities to solicit experiments for these last missions and received over one hundred proposals or inquiries.

The competition to build the spacecraft and cameras was won by the Boeing Company as the prime contractor, supported by two major subcontractors, RCA and Eastman Kodak. Langley's program manager, Clifford Nelson, put together a superb team to oversee the program; many years later, when NASA management called for a review of lessons learned from all the completed programs, Lunar Orbiter was judged the best managed. If for some reason it had not been successful, the entire Apollo project would have been in jeopardy or, at the very least, delayed beyond the date President Kennedy had called for. Lunar Orbiter was successful far beyond our hopes based on our experience with Ranger and Surveyor. *Lunar Orbiter 1,* which flew in August 1966, did not perform completely to specifications, but it returned a total of 422 medium and high resolution photographs of potential lunar equatorial landing sites as well as some photographs of the Moon's farside. After correction of the problem that degraded some of the first mission's photographs, *Orbiter 2* and *Orbiter 3* were so effective that all the Apollo landing site photographic requirements were completed; the engineers and mission planners had enough photographs in hand to permit detailed landing site analysis, and they released the final two spacecraft for science and site selection for potential post-Apollo missions. (The last three Lunar Orbiters were eventually canceled, and the experiments solicited for those missions were put on the shelf to be resurrected later.)

The first three spacecraft had concentrated primarily on photographing the nearside equatorial zone, where the upcoming Apollo landing sites would be. *Lunar Orbiter 4* expanded the coverage on the nearside, including many of our high priority post-Apollo exploration sites. The final mission, *Lunar Orbiter 5,* completed the coverage of the poorly known farside. By the time *Lunar Orbiter 5* snapped its last picture, the five Lunar Orbiters had sent back 1,950 pictures of the Moon covering most of the lunar surface, nearside and farside. The resolution of these photographs ranged from approximately sixty-five meters to five hundred meters, although much higher resolution photographs of the potential Apollo landing sites were taken on the first three missions. To obtain this higher resolution (two meters), the first three missions took their photographs at lower orbital altitudes than the final two.

Thus Lunar Orbiter equaled the best Earth-based photographs, and it bettered many of them by a factor of 250. Only a small area of the Moon was covered by the high resolution photographs, but the coverage had been judiciously distributed by the planning teams. An added benefit was that by closely tracking the spacecraft's orbits, we were able to map the Moon's gravity field at a resolution not achievable from Earth.

Both the Falmouth and Santa Cruz summer conferences devoted considerable thought to recommending experiments that could be done in lunar orbit to complement the study of the Moon from the lunar surface as part of the comprehensive, post-Apollo exploration program. In 1964 and 1965 Peter Badgley had attempted to interest NASA management in a remote sensing program to be conducted in Earth and lunar orbit, and eventually a program titled Lunar Mapping and Survey System was initiated.[1] This program, designed to use Apollo hardware, was canceled in early 1968 in a cost-cutting move.

But the recommendations from the summer conferences did not die. In March 1968, ignoring the just announced program termination, Sam Phillips sent a memo to Bob Gilruth requesting that MSC look into providing scientific and operational photography during the landing missions.[2] With planning proceeding for the final missions, and following up on the Phillips's request, Lee Scherer sent Bill Hess a memo in early May 1968 asking that MSC begin to study how to integrate experiments into the command and service module to take advantage of the longer staytime in lunar orbit. Hess agreed, prompting our office to write a memo for Phillips's signature asking MSC to expand the study he had requested in March to identify other orbital experiments that would take advantage of the "overall CSM science potentialities."[3] This memo resulted in MSC's adding $100,000 to its Martin Marietta Apollo Applications Program integration contract and marked the beginning of a program to develop a suite of sensors that would be flown in the CSM.

While this analysis was under way, OSSA dusted off the experiments that had been submitted earlier for Lunar Orbiter and began to assemble the rationale for including different suites of cameras and sensors that could fit into the CSM. George Esenwein, who had been the headquarters project officer for the Apollo command and service module mechanical systems, transferred to our office at this time and was put in charge of the orbital science and photographic team. Floyd Roberson was named program scientist, and David Winterhalter was program engineer. Noel Hinners, at Bellcomm, assigned several members of his staff to work with this team, notably Farouk El Baz and Jim Head, both of whom had played prominent roles in analyzing Lunar Orbiter photographs and recommending targets for photography on *Orbiter 4* and *Orbiter 5.*

As an extension of these studies, Esenwein's team, working with MSC, determined that it would be possible to include in a service module (SM) bay a small subsatellite that could be left in lunar orbit, and an announcement of flight opportunities was released soliciting experiments that could utilize the subsatellite. In April 1969 OSSA and its advisory panels reevaluated the Lunar Orbiter proposals, and the new proposals to place experiments on the subsatellite, and selected a final suite of experiments.[4] In June OMSF directed MSC to proceed with the modifications of the CSM and to procure the experiments. Eventually the science payload carried in the command and service module, including cameras, experiments, and the subsatellite, totaled almost 1,200 pounds. Most of the experiments were housed in one quadrant of the service module in what was named the scientific instrument module (SIM), and a few were carried in the command module (CM).

For the experiments that did not send their data back by telemetry but recorded them on film or in some other form, the film and data would have to be retrieved by the CM pilot during extravehicular activity. After much debate concerning the safety of the CM pilot during the retrieval operations, it was finally agreed to schedule this EVA after leaving lunar orbit, when the astronauts were safely on their way back to Earth. Imagine floating outside your spacecraft somewhere between the Moon and Earth attached by an umbilical cable and a slender wire! The three CM pilots who carried out this risky maneuver would all comment on the strange sensation of seeing the Earth from so far away while floating in space.

Starting with the flight of *Apollo 8* at Christmas 1968, the astronauts began making their contributions to studying the Moon from lunar orbit. Armed with the ever present hand-held Hasselblad cameras, the crew of *Apollo 8* and all the crews that followed (except *Apollo 9,* which remained in Earth orbit) took pictures of the Moon from various altitudes above the lunar surface. Many of the photographs taken during the early missions were meant to improve our understanding of future landing sites by augmenting the Lunar Orbiter photographs. *Apollo 12,* as an example, took 142 multispectral photographs of the designated *Apollo 13* landing site, Fra Mauro, and other equatorial sites. These photographs were used to help decipher the geology and to improve the productivity of the astronauts after they landed by identifying sampling sites that probably had different mineralogical compositions. After *Apollo 13*'s failure, Fra Mauro became the landing site for *Apollo 14,* and the information obtained from the multispectral photography helped, in a small way, in planning the *Apollo 14* surface traverses.[5]

Apollo 14 carried out a variety of experiments, including photography, while on the way to the Moon, in lunar orbit, and on the return to Earth. Three types of cameras were used: a 16 mm data acquisition camera, Hasselblads, and the Hycon lunar topographic camera. (The Hycon malfunctioned during the mission, but almost two hundred usable photographs were recovered.) These experiments included measurements of gegenschein and heiligenschein (rather arcane observations, the former possibly related to Earth-Moon-Sun libration points[6] and the latter related to reflected light, which had potential application for the interpretation of the Moon's fine-scale surface roughness). An S-band transponder experiment provided new information on the Moon's nearside gravity field by permitting close tracking of the CSM's orbits and a bistatic radar investigation that yielded information on the lunar crust.[7] The final missions, *Apollos 15, 16, and 17,* had much more extensive orbital science payloads than any of the previous missions.[8]

Because I was not closely involved with developing the experiments carried in lunar orbit, I will not further describe them or their principal investigators, but for completeness in covering the scientific results of Apollo, in chapter 13 I briefly discuss the scientific information returned from some of the experiments.

On to the Moon:
Science Becomes the Focus

On July 16, 1969, along with a multitude of other sightseers (local Civil Defense officials would later estimate one million), my family and I were on hand to watch the launch of *Apollo 11*. Our Winnebago camper was parked on the shoulder of U.S. Route 1 about five miles north of Kennedy Space Center and the launch site. We had picked our viewing point the night before, feeling lucky to find a spot so close. It had been a madhouse trying to drive near the Cape; no one seemed to care about following normal rules of the road as cars and campers vied for spots and parked wherever they pleased. Local and state police tried to maintain some order, but it was a hopeless job. In the early morning, as launch time approached, we climbed on the roof of our camper to get an unobstructed view, meanwhile listening on the radio to John "Jack" King, "the voice of Apollo," count down the final seconds.

Old Glory was flying everywhere, and the crowd was in a party mood. The countdown proceeded smoothly, and at 8:32 A.M. the Saturn rocket lifted off accompanied by loud cheers and many teary eyes, mine included. Beyond a doubt our hearts went with the crew of *Apollo 11*. This was the second *Saturn V* launch I had witnessed, but I still wasn't prepared for the enormous noise and low-frequency reverberations that reached us, even at this distance, in the minute after the Saturn cleared the launch tower. We watched for several minutes as it disappeared to the east, leaving behind a huge plume of white smoke, then we went inside, finished breakfast, and talked about what we had just seen. My sons, only eight and eleven at the time, still vividly recall the excitement of that morning. I was in a hurry to leave because I was due back in Washington in a few days, but we were forced to wait almost an hour before the traffic jam began to move and we were back on the road. *Apollo 11* was on its way to the Moon with the first science payloads that men would place on another body in our solar system. If all went as scheduled, Neil Armstrong and Buzz Aldrin would have the honor of making the first direct, close-up studies of how the Moon's surface looked and how it felt to walk on the Moon in one-sixth gravity. After the landing and takeoff from the Moon, Mike Collins, the command module pilot, would be waiting in lunar orbit to rendezvous with the lunar module, ready to lower his orbit if the ascent stage did not perform as well as planned.

Four nights after the launch, in anticipation of the landing, the Voice of America (VOA) had assembled a team to report on this once in a lifetime adventure for its worldwide audience. Several NASA colleagues, Merle Waugh, John Hammersmith, William Land, and I, were in the Washington studios as "color commentators" to back up the VOA reporters led by Rhett Turner, who would be reporting from the Manned Spacecraft Center in Houston. We listened anxiously, just like millions of others around the globe, to the exchange between the capsule communicator (CapCom) Charlie Duke and Armstrong and Aldrin in the *Eagle* as they went through the final maneuvers to land the LM. The excitement of those last few minutes, heightened by the crew's difficulties in selecting their landing site with alarms ringing in their ears and their fuel supply nearing exhaustion, made Armstrong's announcement "Houston, Tranquility Base here, the *Eagle* has landed," almost anticlimactic. We could hear the cheering in the Mission Control Room through Rhett's microphone, and we in VOA's Washington studio were yelling and pounding each other on the back too. Although we had worked for years to help achieve this moment, it seemed incredible that we were successful on the first try.

We were primed to discuss the mission in great detail, but as the night unfolded only a few questions were directed our way, and I was never called on to demonstrate my vast insight into things lunar. VOA wasn't about to share the limelight on this historic occasion. I did, however, receive a card from some friends in Colombia who said they had heard me on VOA. They told me how proud they were of *Apollo 11*'s success and congratulated me on being part of the program. I wondered if some of my former colleagues remembered their skepticism six years earlier when I decided to leave Mobil and join NASA. I certainly did not regret the decision. Our great hopes to follow Apollo with extensive exploration and lunar bases now seemed remote, but important work still lay ahead to make each succeeding mission more scientifically productive.

As the scheduled launch date for *Apollo 11* drew closer, NASA management became more and more cautious and conservative. This was especially evident at MSC, where caution was the trademark, but even at NASA headquarters one could sense growing concern about the many uncertainties and dangers that simulations and planning could not make go away. Mueller's decision to go to "all up testing" had eliminated several test flights that would have provided additional experience, but it was too late to go back and build confidence any further than where we were in July 1969. The only alternative was to schedule a conservative mission profile leaving as much margin for error as possible.

The Early Apollo Scientific Experiments Package (EASEP) that Armstrong and Aldrin would carry on their flight, described in chapter 7, did not represent a complete Apollo Lunar Surface Experiments Package (ALSEP), since both headquarters and MSC feared that the tasks originally planned would be too demanding. EASEP included a solar-powered seismometer and an additional experiment, the Laser Ranging Retro-Reflector (LRRR). The Swiss-sponsored Solar Wind Composition collector would also be deployed, but its scientific value would be degraded because of the short time it would be exposed to the solar wind. Sample collection and photography were scheduled in connection with the crew's geological study, but they were also reduced in scope from the original plans.

Before the launch, word of changes had reached Congress, some of whose members were already chafing at the expense of the program. These changes had raised questions about the cost of removing the planned equipment from the *Apollo 11* mission. On March 13, 1969, just four months before *Apollo 11*'s scheduled launch, the House Subcommittee on Space Science and Applications held a hearing at which a number of questions were asked about the last-minute science payload changes. Chairman Joseph E. Karth (D-Minn.) asked, "Can we put in the record why the ALSEP is not flying on the Apollo trip as originally planned?"[1]

Our office responded four days later with the following explanation:

The goal of the first Apollo mission to the lunar surface is the successful landing and safe return to Earth of the astronauts. The primary objective of the mission is to prove the Apollo system-launch vehicle, spacecraft, spacesuits, men, the tracking network, the operational techniques.

The first landing mission represents a large step from orbital operations.

The descent, landing, extravehicular activity (EVA) and ascent from the lunar surface are new operations in a new environment. Our Gemini EVA experience showed that a methodical increase in task complexity was necessary in order to understand and operate in the zero g space environment. The 1/6 g lunar surface environment will be a new experience. We cannot simulate it completely on Earth. We find, for example, that we simply do not have as much metabolic data as we would like in order to predict with high confidence, rates in a 1/6 g environment. Only educated guesses are possible on the difficulties the astronaut will have in maneuvering on the surface or the time it will take him to accomplish assigned tasks.

Until recently, the first mission plan called for two periods on the lunar surface (EVAs). During the second period, the crewmen would deploy the Apollo Lunar Surface Experiments Package (ALSEP). This would take place immediately prior to lunar ascent and rendezvous. Because of biomedical unknowns, we are concerned with the degree to which the second EVA would fatigue the crew and adversely affect their performance during the critical ascent and rendezvous phases of the mission.

After extensive review and evaluation, we reached the decision not to have a second EVA on the first landing mission. The ALSEP will be deferred to the second mission. We will make every effort on the first landing to obtain data leading to a firm assessment of the astronaut's capabilities and limitations on the lunar surface with a view toward increasing, on subsequent landings, the percentage of EVA time available for scientific investigations. Deployment of the ALSEP on the second mission is planned as a primary objective.

Our answers to other questions raised by the subcommittee included an estimate of $5 million to modify the ALSEP seismometer to the EASEP configuration. (This number differs from the contract cost of $3.7 million discussed in chapter 7 because it includes other costs associated with the EASEP, such as integration and training, that were not part of the Bendix contract.)

Left out of the response was another concern, the performance of the LM during the first landing and takeoff on the Moon. Although the LM had performed well on *Apollo 9*'s Earth orbital flight and *Apollo 10*'s close approach to the Moon's surface, leaks in its propellant tank had only recently been fixed. With only two LM test flights under our belts, NASA management was still concerned about this problem. Our office was understandably chagrined at the changes in the timeline and the science payload, but this turn of events lent even greater importance to ensuring that the science planned for the next landings was not compromised.

Another interesting exchange before a Senate committee took place shortly after the House subcommittee hearings. Homer Newell and John Naugle appeared before the Senate Committee on Aeronautical and Space Sciences on May 1. During the questioning, Senator Carl T. Curtis (R-Neb.) asked Newell and Naugle if knowledge gained from our completed space missions had changed previous beliefs. Both Newell and Naugle said yes, and Newell went on to provide a surprising example. He said that the "mascons" discovered by tracking the Lunar Orbiter flight paths (concentrations of high density material below the surface of the lunar seas that might indicate large meteor impacts) "give rise to some of the speculation that maybe at one time these areas were actually oceans or seas and [that] sediments from these oceans or seas is what filled those holes." You won't find these speculations in chapter 2, although many thought there was a chance that some water had been present on the Moon at one time. The theory that the Moon once had oceans was not supported by any prominent theorists of the day, and if the large impact craters had been filled with sediment of some kind they would have been deficiencies of mass, not mass concentrations. For a crater to be a "mascon," the fill had to be some unusually dense material. Even senior NASA managers had a hard time keeping up with changing theories as new information was gathered and analyzed by more and more students of the Moon.

One week before *Apollo 11* lifted off, Sam Phillips issued a new Apollo program directive (APD) detailing a total of ten lunar landing missions.[2] The first landing was designated a G mission with the characteristics noted above, and the next four were called H missions. The H missions were designed around two EVAs, surface staytimes of up to thirty-two hours, and our old reliable payload of some 250 pounds. The final five missions, *Apollo 16* through *Apollo 20*, were called J missions. Although the APD did not specify any science payload numbers, it stated that both the lunar module and the command and service module would be improved to permit longer staytimes. We anticipated that the LM would be able to carry additional descent propellant, which would translate in part to an ability to carry larger science payloads. We still held out hope in 1969 for flights beyond *Apollo 20*, but realistically we would have to extract as much science as possible from these ten missions. It was not exactly what we had planned for in 1964 and 1965, but we expected the J missions to be far better than the original Apollo plans. An interesting statement in the APD was that the constant-volume space suit would be available for the J missions. This never came to pass, and if such suits had been used they probably would have had little effect on the productivity of the J mission EVAs. LM and CSM consumables became the limiting factors, not the astronauts' metabolic rates.

As we had simulated at Martin Marietta in 1964 and 1965 in case of an abort after touchdown, the crew of *Apollo 11* first used their eyes to describe the lunar scene and took a few photographs before leaving the LM. One other piece of data collected was a movie of the landing site filmed from Aldrin's window as Armstrong maneuvered for the landing. Not much scientific use was made of this movie because of its limited view of the surface, but you could see how the Moon's surface layer was disturbed by the exhaust of the LM descent engine, with the fine-grained particles shooting rapidly away from below the LM in a fuzzy blur. These pictures confirmed that the lunar surface reacted as predicted to the LM exhaust and helped ease concerns about future LM landings. Peering out his small window, Armstrong provided the first descriptions of the surface, and Armstrong and Aldrin took pictures with the Hasselblad camera and described what they could see from their windows. Their observations added to the overall understanding of the landing site but did not reveal precisely where they had landed.

Whether Armstrong or Aldrin would have the honor of being the first human to stand on another celestial body had been decided long before *Apollo 11* was launched. The initial timelines, circulated almost a year earlier, had indicated that Aldrin would be the first out. As planning for the mission matured, however, it became evident that the LM commander, Armstrong, would be in the best position inside the LM to perform this historic first, seniority notwithstanding. From a science standpoint it really didn't make any difference who would be first on the surface, but for Aldrin the decision was obviously a disappointment, and it continued to trouble him years later. Usually few people remember who was the second to do something; however, both Armstrong's and Aldrin's names are synonymous with the first Moon landing. Through the years Aldrin has received his deserved recognition, but he is not quite as famous as if he had been the first to touch the Moon.

It took some time for Armstrong and Aldrin to secure the LM and get it ready for a quick takeoff, if necessary. After landing and preparing for an emergency takeoff, the timeline scheduled a meal followed by a sleep period. The astronauts, understandably excited and not sure how long they would be permitted to stay on the Moon, asked Mission Control to skip the sleep period and immediately begin preparing for their EVA. Receiving approval, they donned their space suits, and a little under seven hours after they landed Aldrin opened the hatch. Armstrong squeezed through and bounced down the ladder (without seeing any exploding "Gold dust").

His descent and first steps on the Moon were recorded for all the world by a television camera attached to the landing stage, which he activated from the top of the ladder. This camera, built by the Westinghouse Aerospace Division, had been the subject of much debate. Could we afford the weight (about ten pounds) and the complications of deployment, since we knew the quality of the pictures would be poor? I was for not carrying it, especially when we were discussing whether to include the ALSEP because of weight and EVA time concerns. But once it was decided to eliminate the ALSEP, the question became moot from a science perspective. The "let's carry it" side won the day, and it turned out to be a valuable tool both for public relations and for science. We used the TV pictures, in spite of their poor resolution, to help reconstruct the astronauts' movements and plot the geology. Some senior NASA managers complained during the mission about the poor quality of the pictures, but by then it was too late. (The poor picture quality was caused not by any Westinghouse design deficiencies but by the NASA specifications, dictated by weight and power constraints and antenna performance.)

After examining the LM and reporting its status, Armstrong began describing the scene around him and his impressions of the lunar surface. Then he took a few photographs. He collected the contingency sample and put it in a pocket of his space suit, and he was soon joined by Aldrin to complete their carefully choreographed EVA timeline. At this point Armstrong removed the TV camera from the LM and set it up some sixty feet to the northwest, providing a limited view of the landing site and of the astronauts' movements as they went about their EVA. From this time until Armstrong and Aldrin reentered the LM, they performed all their tasks as planned. I won't go into detail on what they accomplished; references listed in the bibliography describe these activities in great detail. Both astronauts performed all their scientific assignments better than expected under extraordinary conditions. One might think that the first men to land on the Moon might not have their minds completely on the scientific tasks before them. One might expect them to be thinking about the upcoming liftoff, a maneuver never before attempted, which their survival depended on. Armstrong and Aldrin seemed to put such concerns out of their minds. They appeared to be completely absorbed in deploying the experiments, sampling, and describing what they were seeing and doing.

Aldrin placed the EASEP, the last-minute replacement for the ALSEP, on the surface about sixty-five feet south of the LM and in the same general area as the Laser Ranging Retro-Reflector. He had no trouble unfolding the solar panels and erecting the radio antenna, and once set up the experiment turned on automatically. Back on Earth, signals were received almost immediately, relayed to Houston from the NASA Manned Space Flight Network. We knew it was working because the seismometer recorded Aldrin's footsteps as he walked nearby, but we hadn't expected to receive so many signals.

The MSC and Bendix engineers manning the EASEP console in the Science Support Room (SSR) soon began to see a problem. The temperature of the seis-mometer package was rising faster than expected. It took some time to arrive at a probable cause, but they finally decided that dirt and dust were covering some of the surface, reducing its ability to reflect heat. Both Armstrong and Aldrin had commented on how far the soil would fly when they walked, as well as on how dirty their suits got during the EVA. While deploying the EASEP they had completely circled the experiments, so it was logical that some soil had coated the surfaces. Also, based on Aldrin's comments, as we continued to track the rising temperature after their takeoff, it appeared he had placed the EASEP experiments closer to the LM than requested. We assumed that dust thrown up during the takeoff had also been deposited on the experiment surfaces. We kept our fingers crossed that the soil would not overheat the seismometer and had not obscured the small corner reflectors of the LRRR, making it difficult to bounce laser beams back to Earth.

These eventualities didn't come to pass; the seismometer survived the rest of that lunar day (fourteen Earth days) and the following lunar night and came back on line for seven more days when the solar panels saw the sun again. The seismometer recorded several interesting events during its short lifetime, in-cluding the shocks of the astronauts' backpacks hitting the lunar surface when they were thrown from the LM and the small "moonquake" when the ascent stage lifted off. Based on this performance, we could anticipate that the seismometers of the same design scheduled for the full ALSEP deployments would provide even more information during their much longer lifetimes.

In addition to still photographs, movies, and the Solar Wind Composition collector foil, a total of some forty-seven pounds of individual rocks, soil, drive-tube cores, and the contingency sample, all neatly packaged, finally found their way to MSC, where the staff at the Lunar Receiving Laboratory, and eventually the sample analysis principal investigators, eagerly awaited them. On the recovery aircraft carrier, the USS *Hornet,* the samples were divided into two batches and flown to Ellington Air Force Base in separate aircraft to ensure that some samples would survive in case one plane was lost at sea. There was always the chance we might not get back again to collect more samples. From Ellington, they were carried to the LRL.

The astronauts, wearing isolation garments that they donned in the CSM while awaiting recovery and transport to the *Hornet*'s deck, were immediately sequestered in a specially designed trailer lest they contaminate those around them with some deadly unknown virus. After the *Hornet* arrived at Hawaii, they too were flown back to MSC in their trailer along with two volunteer MSC doctors, to begin their one-month quarantine.

The samples, which had arrived before the astronauts, were carefully opened in the LRL, inventoried, and briefly described. In the meantime we were monitoring the signals sent back by the passive seismic experiment and attempting to find the LRRR that the astronauts had left behind. This latter operation was not as easy as we expected, since the exact location of the landing site was not immediately known. Mike Collins had attempted unsuccessfully to locate the LM from orbit using the command module sextant. After analyzing the flight data and the returned photographs, we passed our best estimate to the LRRR PIs, and the LRRR was found on August 1, 1969, by the Lick Observatory in California.

On August 23, 1969, one month after *Apollo 11* splashed down and the date when the astronauts were released from quarantine, George Mueller forwarded a memo to Clare Farley, James Webb's executive officer, to be included in the report being sent to the president summarizing the results of man's first foray to the Moon.[3] In his memo, drafted in part by our office, he described the initial scientific results of *Apollo 11* and summarized the program adjustments that would be made as a result of the mission. Included with the memo was a preliminary traverse map compiled by Gerry Schaber and Ray Batson of the United States Geological Survey using tapes from the lunar module's television camera, photographs taken by the astronauts, and educated guesses based on what the astronauts reported from the Moon. The map sent to the White House had been further updated during the astronauts' debriefings while they were still in quarantine. By this time photographs of the astronauts on the Moon and a few photographs of "Moon rocks" had circulated in all the newspapers and some magazines, so Mueller didn't include any photographs of the astronauts with his memo, but he did include a photo of one of the returned samples. The Schaber-Batson map had just been completed and represented new information not yet made public, tying together everything the astronauts had done during their brief stay.

Short and to the point (five pages plus attachments), Mueller's memo provided an initial age dating of one sample (3.1 billion years), compared the chemical and mineralogical content of a few samples with that of the Earth, and offered the conclusion that the Earth and the Moon probably were formed "from the same whirling cloud" some 4.5 billion years ago. (It wasn't clear where that comparison came from, but it wasn't too bad a description if you agreed with the conclusion.) He also briefly discussed some results from the passive seismometer and LRRR; the latter experiment permitted the measurement of the Earth-Moon distance to within twelve feet as opposed to the best previous accuracy of about two thousand feet. (The accuracy of a few inches predicted in chapter 7 would come only after several years of ranging from three or more stations.) The last sentence we added to the memo was, we hoped, a thinly veiled plea to the White House to the keep missions going: "The indications thus far are that the Moon is a celestial body with complex structure, geology, and chemical history that may take considerable effort to unravel."

Mueller's attachment summarizing planned program adjustments had an important effect on the subsequent missions. With the lunar landing mandate successfully completed, Mueller now proposed to slow the pace of the missions from one launch every two and a half months to one every four months. He stated that this not only would save money but would allow us to "increase mission flexibility and scientific return in later missions." This was a welcome change to those of us planning the science and to the staffs at MSC and KSC, who had been working around the clock to support the shorter schedule. This would, we hoped, allow us to factor in some of the results of the previous missions while developing the objectives for each succeeding one and to alter the science payload and astronaut training accordingly. To a large degree we were able to do this on the last three J missions.

With the flight of *Apollo 11* successfully concluded, General Phillips relinquished his position as Apollo program director and returned to the Air Force. He was replaced by Rocco Petrone, who until this new assignment had been director of launch operations at KSC. Rocco, a West Point graduate, was a large man. He had been a backup tackle on two of Coach Red Blaik's most famous Army football teams of the 1940s, which won thirty straight games before being defeated by Columbia in 1947, my freshman year. The teams featured "Doc" Blanchard charging up the middle or Glen Davis scampering around the end, at times behind the broad back of Rocco Petrone. He was listed in the game programs of the time as six feet one and 202 pounds; in the 1940s these were not intimidating numbers for a tackle, but he wasn't exactly small. In 1969 he was a little more imposing, perhaps with a few more pounds than he carried in his playing days.

I don't have many recollections of specific meetings with Sam Phillips, but I do remember calm, quiet, efficient status reviews that moved along quickly, with Phillips clearly in command—a management style much like George Mueller's. Meetings with Rocco were different. He came to Washington with a reputation as a hard-nosed, hard-driving manager with his record at KSC—all Saturns launched successfully—a testimony to his management skills and his team's ability. He had succeeded in what must have been a difficult environment under the early tutelage of the German-trained rocket scientists assembled by Wernher von Braun and KSC director Kurt Debus, both known to be sticklers for detail and perfect performance.

Rocco was the only senior manager I worked with who truly had a photographic memory. If you gave him a "fact" related to your program during a briefing, woe unto you if you changed anything a week, month, or year later. Rocco would catch or challenge you, and he was almost always right. Rocco's meetings were a little more lively than Phillips's, especially if there were discussions of delays or unexpected changes. He was never shy about showing his displeasure, and it was reinforced by his imposing frame. Conference calls between Rocco and the NASA centers were always interesting. Usually they were arranged to discuss some critical problem, so by their very nature they were bound to be contentious. As we listened to Rocco asking questions in his distinctive high-pitched, singsong voice, we could visualize the speakers at the other end of the line squirming as they tried to justify some earlier position that he didn't agree with. Rocco soon became our strong right arm and a defender of lunar science. Once he was convinced of the correctness of a scientific position, we seldom lost any ensuing argument with MSC. After Rocco's arrival we really buckled down to expand and improve the science on the last three missions.

Flight readiness reviews (FRRs) were another area where Rocco ran a taut ship. Hosted by KSC, they were the final review, held about one week before a scheduled launch. Chaired by Chester "Chet" Lee, Rocco's Apollo mission director, they usually lasted one full day. There were representatives from all the NASA centers involved in the launch as well as the contractors and the required Department of Defense participants—a cast of hundreds. Every aspect of the mission from prelaunch preparation to splashdown and recovery was discussed in detail and checked off as being ready if it passed the rigorous review. Action items or deficiencies recorded during earlier mission reviews were carefully analyzed to be sure they had been properly attended to. This process might result in long debates, followed by documentation to prove problems had been resolved. Any items still open after the FRR were subject to a final review and structured sign-off before launch. Here is where Rocco's photographic memory was put to the test. He would recall the smallest detail and ask penetrating questions. If the presenter could not answer to his satisfaction, someone had to leave the room and gather the missing information.

FRR attendance was carefully controlled. NASA senior management was seated at the front of the room, along with at least one of the astronauts who would be on the crew or serve as backup crew for the launch under review. Briefers with their supporters scurried in and out as called for by the agenda. For the J missions, I was entitled to a seat at the back of the room to take notes and perhaps pass on a discreet question for Chet Lee or Lee Scherer to ask. But the FRRs tended to be a one-man show, with Rocco calling the shots and the other senior managers like James McDivitt, Deke Slayton, and Al Shepard recognizing his mastery of the occasion. Everyone knew Rocco's reputation for detail, and no facts or concerns were held back. We all understood that the lives of the astronauts seated in the room with us could be in jeopardy if the smallest problem went undetected or unsolved.

Hangar S became a kind of science headquarters at KSC as we approached the Apollo lunar missions. It was a little seedy looking on the outside—the paint was peeling and the large S was barely readable—but the inside was a high-tech workshop. As the name indicated, it was formerly a hangar at Cape Canaveral Air Station, but it now functioned as an important facility at KSC where final preparations and checks were carried out for all the experiments. Mock-ups of the LM and CSM were maintained in the hangar and used for stowage checks and simulations, which became increasingly complex for the missions following Apollo 11. The crews would spend more and more time at KSC as they neared the launch date, so it was important to have a place where they could stay up to date on any changes that might involve the experiments.

Flight experiments were sent to KSC from contractors around the country. KSC engineers would receive the flight hardware and store it in a clean room in another building near hangar S where final checks would be made to ensure that nothing had been damaged during shipping. Contractors building the experiments and equipment did their own inspections before the items left their plants, but the final checks were done at KSC. Nothing was loaded on the LM or CSM if it had not undergone a rigorous preflight inspection. Once it passed this inspection, it would be taken to the Vertical Assembly Building to be stowed.

Since ALSEP was the major science payload after the flight of Apollo 11, it received the most attention. It was carefully unpacked in the clean room, and each experiment was set up to check cable connections and any unique fasteners, thermal blankets, or other apparatus that might give the astronauts trouble during lunar deployment. Chuck Weatherred, the Bendix ALSEP manager, recalled an important exchange as he helped the KSC team prepare for the launch of a "dummy" ALSEP on Apollo 10, scheduled to fly to the Moon but not land. Peter Conrad and Richard Gordon, the Apollo 12 crew, came into the clean room to watch the processing of the package that would simulate the weight and center of gravity of the ALSEP so that the MSC flight dynamacists could calculate how the spacecraft would react to various maneuvers during the mission. Although they had visited Bendix and seen their ALSEP in the final stages of manufacture, they knew their training schedules did not call for them to have any direct interaction with it until they were on the lunar surface. Conrad asked Chuck if they could participate in the final checkout before their ALSEP was stowed for the journey to the Moon. Chuck thought that was a great idea and said he would get approval from MSC, but Don Wiseman, his MSC contract manager, turned the request down. MSC didn't want the astronauts fooling with the flight hardware before they deployed it on the Moon.

After several appeals and backing by the astronauts, that decision was reversed, and all crews starting with *Apollo 11* were permitted to work with the flight hardware at KSC before it was finally stowed for the trip to the Moon. It was perhaps a small victory, but I feel sure it made the crews more confident that they would not confront any surprises. "Murphy's Law" says anything that can go wrong will go wrong. No matter how closely you monitor the manufacture of such a complex set of equipment as ALSEP, minor changes not reflected in the simulation hardware or documentation (someone's last-minute bright idea) can creep into the design and could cause complications 238,000 miles away. We had few such problems with the science payloads, in part because we worked hard to be sure the astronauts were always in the loop.

At the same time that we were savoring the success of *Apollo 11*, the National Academy of Sciences' Space Science Board was conducting another summer study, once again at Woods Hole. The study was chaired by Harry Hess of Princeton University, who had also led the 1965 summer study held in conjunction with the Falmouth conference. Harry was a strong advocate of manned and unmanned lunar exploration, and his position at the Academy as well as his overall reputation in the scientific community lent great weight to our Apollo science planning. Harry's objective for the study was to capitalize on *Apollo 11*'s success and lend support to those of us arguing with the administration and Congress to use the remaining Apollo hardware to carry out more missions and missions with ever increasing exploration potential.

Immediately after *Apollo 11*'s return, some leading decision makers in and out of Congress, who will remain unnamed, had been quick to propose ending lunar exploration and spending the money saved on various social programs back on Earth. These discordant voices motivated Hess to quickly call for the study. I attended the meeting with Don Wise, who had joined our office from Franklin and Marshall University to be Lee Scherer's deputy. We made several presentations based on our ongoing efforts for the J missions, pointing out the potential for enhancing the science return. We also reviewed the recommendations of the Santa Cruz conference and the "Lunar Exploration Plan" we had disseminated at the end of 1968. This summer study provided a new opportunity to resurrect some of our old plans for long-duration missions that we were forced to abandon in 1968 for lack of interest by Congress and the administration. Along with many other participants in the Apollo program, I strongly supported Harry's views that we must make the case to take advantage of this opportunity—to squeeze as much science as possible from the Apollo program. After all, the major expenditures had already been made; using all the hardware, and doing it more efficiently, would entail adding only a small fraction to the total spent to date for the new science payloads and mission operations.

Tragedy struck the study on the first day, August 25, 1969. During the morning coffee break Harry complained of chest pains and left to see a doctor. He never returned. We were told he died peacefully at the doctor's office. This, of course, spread a pall over our meeting. We had lost an irreplaceable leader whose vision had been, since the earliest days, a major force in our efforts to bring good science to the Apollo program. Only a few special people, including Ralph Baldwin, Harold Urey, and Gene Shoemaker, can lay claim to being fathers of lunar exploration, and Harry Hess belongs in that company. We continued our deliberations under a new chairman, Bill Rubey, the newly appointed director of the Lunar Science Institute, and then issued our report.[4] A case was made to support the launch of the nine missions still being planned at the time and to continue additional missions through 1975. The study concluded: "The decision concerning the nature of the lunar exploration program after the mid-1970s will hinge on the national commitment to manned space flight and on the significance of the scientific discoveries that emerge in the next few years."

While this report was in press, those of us advocating more Apollo science received another blow. Bill Hess resigned from his position as director of science and applications at MSC; he finally got tired of bucking the entrenched antiscience interests there. Tony Calio, who had earlier worked with us on Foster's staff, took Hess's place. When Tony left our office to go to MSC, we gave him a going-away party in Washington, wished him success in taking on such a difficult position, and looked forward to having someone at MSC who would be receptive to our interests. At the time, we didn't know his appointment would adversely affect our relationship with MSC, but within weeks it became apparent. Tony quickly adopted the MSC line, and our relationship with MSC regressed to where it had been two years earlier. He became hard to reach by phone, and when we did get through he ignored most of our suggestions. He also developed an intense dislike for the staff at USGS. I never fully understood the reason for this antagonism—perhaps it was a holdover of earlier disputes between USGS and some of the staff he inherited. But this undermined USGS's ability to support the upcoming missions for which members of the Field Geology Team had an ever increasing responsibility. It was only through their close relationship with the astronauts and others in the astronaut office that they were able to influence the geology content of the missions.

Returning to the remaining missions, *Apollo 11*'s success and a ringing endorsement from the National Academy of Sciences energized many in the science community to propose exciting new experiments for the remaining missions as we geared up to take advantage of a relaxation in some of the mission constraints. Until *Apollo 11* returned safely, every Apollo engineer and system and subsystem manager was holding a little in reserve just in case it was needed. A little extra weight, a little extra available propulsion, a little extra performance margin. Slowly, with the help of the Bellcommers, these margins were identified and translated into increased science payload and more operating flexibility.

The Schaber-Batson map was the first attempt, other than during simulations, to reconstruct in near real time what was happening on the Moon. Although during the *Apollo 11* mission there was no direct exchange between scientists on Earth and the astronauts, based on our Flagstaff simulations we could see how this could be done effectively for the later missions. For *Apollo 12* and the remaining four missions we tracked the astronauts in real time and had an up-to-the-minute map of their progress in the SSR. We coordinated our tracking with the flight controllers and medical staff monitoring the astronauts' performance to ensure that their traverses would not overextend their life-support expendables. This monitoring was especially valuable during the last three missions, when the astronauts were often far from the LM and we had to be sure they had enough life support reserve to walk back if the lunar roving vehicle failed. For the science team it had another important aspect: it allowed us to relay suggestions for modifying the astronauts' activities through the CapCom as they reported their findings and, at times, changed the timelines on their own initiative.

In September 1969 we advertised the opportunity to propose new experiments for the J missions that would utilize the LRV and the longer staytimes. This announcement, while directed primarily to missions 16 through 20, indicated that proposals to perform simple experiments on flights earlier than *Apollo 16* would also be accepted.[5] Perhaps the most ambitious aspect of this announcement was our optimism about where we would be permitted to target landing sites for the flights that would follow the initial landings. Scientifically exciting sites recommended by the Group for Lunar Exploration Planning (GLEP), such as the central peaks of Copernicus and the rim of Tycho, were included as candidates in the announcement so that proposers could consider their unique characteristics for their experiments.

With the arrival of Tony Calio and the immediate change (for the worse) in climate at MSC in regard to science, we began to lobby Rocco Petrone to push MSC to modify management's responsibilities for science in the hopes that this would improve our working relationship. He talked to Jim McDivitt about making some changes. At the end of October 1969 our office originated a memo for Petrone's signature formalizing these suggestions. The opening sentence, underlined, stated, "*I think we have a problem in the management of the science program which warrants immediate action.*"[6]

McDivitt responded two weeks later and gave us half a loaf.[7] He moved the design, development, testing, and delivery of approved Apollo experiments from Calio's office, the Science and Applications Directorate, to the Engineering and Development Directorate, managed by Max Faget. (We weren't sure if this was a victory.) S&AD would still be in charge of the scientific requirements, science mission operations, postflight data analysis, and interactions with the PIs, but McDivitt promised that his office would strengthen its science oversight. This was encouraging, since Petrone and McDivitt usually agreed on the important aspects of the missions, and science would take center stage for the remaining flights. In spite of these changes, our concerns would soon be echoed by the scientific community.

Through 1970, we were still hoping dual-launch missions might be reinstated, enabling fourteen-day stays on the Moon, and the trade journals of the day continued to write about these plans as if they were approved.[8] In Lee Scherer's office we continued to study an LM shelter and a dual-mode (manned and automated) roving vehicle. Scherer urged Marshall Space Flight Center to complete the preliminary design and promised funding for this work.[9] Meanwhile, preparations continued for the next landing. *Apollo 12*, we hoped, would allow us to accomplish some of the science originally scheduled for *Apollo 11* but at a different mare site, many miles to the west.

Apollo 12 was successfully launched in November 1969 and landed about eight hundred miles west of Tranquility Base at the lunar feature called the

Ocean of Storms, another mare site. If our photo interpretations were correct and the landing site was on an ejecta ray from the crater Copernicus, a few hundred miles to the north, we hoped to return samples of material from deep within the Moon, excavated by the impact that formed this huge crater, some forty miles in diameter. Copernicus is one of the craters you could identify under proper lighting conditions with your ten-power binoculars, just a little west-northwest of the center of the Moon.

Two EVAs were scheduled and carried out, and a full ALSEP was deployed. Peter Conrad and Alan Bean proved to be enthusiastic lunar explorers. Much was made in the press of Pete's laughing, giggling, "cackling," and joking as he went about his tasks, but he and Al performed flawlessly, bringing back some stunning pictures and a wide assortment of lunar rocks. The TV camera, similar to the one carried on *Apollo 11,* was damaged soon after they climbed down from the LM, so we were completely dependent on their oral descriptions to reconstruct where they were and what they were doing. Our simulations at Flagstaff and at other locations once again paid off, and we produced a map of the landing site in the SSR based on their descriptions and dead reckoning of how far they traveled between sampling stations.

In addition to the sample collecting, a major objective of *Apollo 12* was to land near enough to *Surveyor 3* to allow the crew to walk to it and take pictures of the landing site for comparison with the Surveyor TV camera pictures sent back to Earth two and a half years earlier. They would try to bring back pieces of the spacecraft, including the TV camera mirror and scoop, so we could study the effects of thirty months of exposure to the lunar environment. The trajectory engineers in mission control and Pete's piloting skills put the LM right on target, within a few hundred feet of *Surveyor 3*. This demonstration of the ability to land at a precise point on the Moon, as opposed to *Apollo 11*'s overshooting the landing point, eased some of management's concerns as we advocated more difficult future sites. All objectives of the mission were met, and the ALSEP became the first link in the network that the geophysicists had dreamed of for over five years. By the end of their two EVAs, Conrad and Bean had successfully deployed the ALSEP (they encountered a minor difficulty while removing the fuel cask of the radioisotope thermoelectric generator from its stowage on the LM, but deployment proceeded as planned), retrieved pieces from *Surveyor 3*, and collected a wide variety of samples totaling some seventy-five pounds.

While Conrad and Bean were on the lunar surface, Dick Gordon, the CM pilot, was carrying out his tasks. Soon after the others landed he used his sextant to search for the LM on the surface and was successful, even observing the much smaller *Surveyor 3* a short distance away. His primary job was to photograph the Moon from orbit using a Hasselblad and a new camera array called the Multispectral Photography Experiment. The array consisted of four 70 mm Hasselblad cameras with fixed focus, each equipped with a different filter to return photographs in the blue, red, green, and infrared portions of the optical spectrum. This camera array was flown originally on *Apollo 9* with Paul Lowman as PI. (For *Apollo 12,* Alex Goetz of Bellcomm was PI.) Gordon would point the array through one of the CM windows and trigger all four cameras simultaneously every twenty seconds. The major objective was to photograph potential landing sites and, we hoped, use the pictures to extrapolate the re-turned samples to wide areas of the Moon based on spectral differences caused by compositional variation in the lunar soil and rocks. A good concept, but the Moon was not cooperative. When the photographs were developed subtle dif-ferences between the crater Lalande and Mare Nubium were found at only two points. We would have to wait until the J missions, with their more sophisti-cated sensors, to have this exploration technique pay off.

During debriefings of the *Apollo 12* crew we asked why they had moved some of the rocks they sampled before documenting their location with photo-graphs, the preferred technique. Their answer was simple and logical. During their early sampling, they had found that many of the rocks they picked up and had documented were too large to fit into the sample bags. Because they were half buried their full size could not be estimated—they were like "the tip of an iceberg." Rather than waste time photographing samples they could not save, they elected to lift some of the rocks before taking the requested six photo-graphs. As a result of this crew observation, the photo documentation require-ment for the next mission, *Apollo 13,* was reduced to five per documented sample (although that crew never had the opportunity to use the new standard) and continued to be revised, downward for subsequent missions as we better understood the documentation needs for mapping and cataloging the samples in the LRL.[10]

Although Gene Shoemaker was still officially the PI for field geology on this mission, Gordon Swann took over crew training and led the interaction of the Field Geology Team with the crew. (Swann would later be named PI for *Apollo*

14 and *Apollo 15*.) We exercised the crew at the Cinder Lake Crater Field simulation site outside Flagstaff, described in chapter 9, and other sites, and by mission time Swann and his team had established a good relationship with Pete and Al. They had both been good students, and their training carried over to the lunar surface. In addition to Pete's enthusiastic, nontechnical descriptions of what he saw, he and Al also provided a good specific commentary that we easily followed, and the Field Geology Team was able to construct a real-time geologic map of the landing site.

After the mission returned we received a letter from a research physicist at the Atomic Energy Commission's Lawrence Radiation Laboratory in California highly critical of the astronauts' oral descriptions and their apparently poor training. We always responded to letters from the public on any subject. I was assigned to write a letter back for Rocco's signature, and it seemed clear to me that the criticism was based on the press reports of Conrad's voice transmissions, not on the whole transcript.[11] In the response I included some of the astronauts' descriptions not carried by the press, such as the characterization as "granitelike," the various colors they reported, and many other precise descriptions of rock shapes and soil conditions on the lunar surface. I hoped our response was reassuring to this concerned taxpayer. It was meant not to belittle his concerns but to show that this aspect of the missions—the astronauts' geological training—was being seriously pursued so that based on their observations we could extract a vast amount of information from each mission.

Apollo 12 had already gone to the Moon and returned before we were presented with the detailed analyses of the *Apollo 11* lunar samples. This delay was dictated by the quarantine requirements and by an agreement with the sample PIs not to release their findings until a formal conference could be held in January 1970, when all the results would be available.

Two months later, in March 1970, a new solicitation was issued that required scientists wishing to analyze lunar samples to submit, or resubmit, proposals to receive samples returned by *Apollo 14* and subsequent missions. John Pomeroy joined our office at this time to manage the expanded sample analysis program and oversee the operation of the Lunar Receiving Laboratory. By July we had received 383 proposals, including proposals from 175 of the 193 teams (the number had grown from 142) that had analyzed samples from *Apollo 11* and *Apollo 12*. Foreign interest in doing analyses was also growing, and of the 208 new proposals, 95 were from foreign investigators. Gerald "Jerry" Wasserburg, a sample PI from Caltech, writing to administrator Tom Paine in June about his recent trip to Europe, reported that "there is a fantastic amount of enthusiasm by all the scientists who are involved in these different countries, and . . . the foreign press has given them a tremendous amount of coverage. Some individuals, in fact, have become sort of national heroes."[12] As before, almost all the proposals received were accepted, and many of these investigators and their successors still attend the annual conferences at the Lunar and Planetary Institute in Houston.

Before any of the missions, toward the end of 1964 I proposed to NASA management that we study the possibility of commanding the discarded LEM ascent stage to strike the Moon near seismometers that would be placed on the lunar surface by future astronauts.[13] At the time, there was no plan to control the impact point of the ascent stage; if not controlled, it would gradually lose altitude and hit the Moon at some unknown time and place. If we could control the impacts of the LEMs, we would have the equivalent of large explosions that would be recorded by the network of seismometers we hoped would soon be in place. We could not be sure when a moonquake or a meteor might provide an energy source large enough to let us study the Moon's interior. The seismometer packages would have finite lifetimes to record some large natural event; if such events were rare, and if the seismometers malfunctioned, they might not be operating when one occurred. Also, the ascent stage was a rather flimsy, lightweight structure, and I feared its impact might not be recorded if its natural decay from lunar orbit occurred some distance away or perhaps even on the Moon's farside.

We began to explore this idea with MSC and enlisted the support of Frank Press, Bob Kovach, and Maurice Ewing, all members of the seismic teams. It took several years to obtain approval for this maneuver, but by the time *Apollo 12* flew we had an agreement to control the impact point of the ascent stage by using the fuel remaining after rendezvous to make it leave orbit at a planned point. For *Apollo 12* we recorded the astronauts' movements and LM takeoff on the ALSEP seismometer as we had for *Apollo 11*, after which the Moon settled down again and was quiet until the ascent stage hit five hours later, about forty miles away.

We calculated that the impact was equal to setting off an explosive charge with an energy equivalent of about one ton of TNT. The first seismic wave arrived at the *Apollo 12* ALSEP 23.5 seconds after impact, building to a maximum amplitude about seven minutes later, with the total recorded event lasting some fifty minutes. The signal recorded was unlike any seismometer recording observed on Earth after either a manmade or a natural event, especially if one considered the relatively small amount of energy involved. This led to a number of theories about the unusual composition of the Moon's outer layers that might cause such a response. We would have to wait for more information, gathered by the next ALSEPs, before a model of the Moon's interior finally emerged that most geophysicists would agree with. When we described to George Mueller the effect of the LM impact and the unusual response, he said, tongue in cheek, that the large amount of titanium found in the lunar samples suggested the Moon must be a hollow titanium shell—a spacecraft from another galaxy covered with cosmic flotsam and jetsam.

At the end of 1969 Mueller resigned. He had steered the Office of Manned Space Flight, and NASA, to its improbable goal of landing men on the Moon and bringing them safely back to Earth. His management skills have been described by many, and I hope I have given a few insights that will add to an appreciation of those skills. Like Rocco Petrone, he embraced the importance of ensuring that good science be accomplished on the missions. Although I have never been able to ask him why he left NASA, I would not be surprised if a major reason was his frustration at failing to persuade the political powers to approve a long-range plan for continuing manned exploration to the Moon and Mars using the capabilities he and many others had worked so hard to build.

He was replaced by Dale Myers, who had been North American Rockwell's manager for its Apollo spacecraft contract. Dale had survived both the bad times at Rockwell, when the contract was in trouble for many reasons, and the good times starting with the success of *Apollo 8*. It must have been a major culture shock to move from being a contractor who had to bow to his NASA "bosses" to being in charge. But he handled it well, and he had a seasoned team to lean on in his first days. I participated in a number of briefings for him early on, and we hardly skipped a beat as we brought him up to speed on all aspects of the program. He selected Charles Mathews as his principal deputy and Charles Donlan as his technical deputy; both were old NASA hands who could help him understand some of the pitfalls he faced. Eight years later, after we had both left NASA, our professional paths would cross again when Dale was appointed undersecretary of the newly created Department of Energy and I was his acting assistant secretary for conservation and solar energy.

Myers's appointment was only one of several major senior management changes made at this time. Other new blood included George Low, whom Tom Paine, Webb's successor, brought to Washington from MSC to be deputy administrator. All these changes had little effect on the upcoming flights. It did seem, however, that once in Washington Low became more sympathetic to the needs of the scientific community, and he strongly supported the efforts to place a high priority on the scientific returns from the final missions.

Once we had an agreement to control the impact of the LM ascent stage, after *Apollo 11*'s successful mission I proposed deliberately targeting the upper stage (the SIVB) for a lunar impact. This was a lot harder sell than controlling the impact of the LM ascent stage. The SIVB stage, as described in chapter 5, was programmed to deliberately miss the Moon. If it was maneuvered for an impact after placing the CSM and LM on a translunar coasting trajectory, it would arrive at the Moon about the same time the astronauts would be braking into lunar orbit. This was why the original mission rules called for the SIVB to be steered away from the Moon after translunar injection, to avoid any chance that it might interfere with the CSM and LM.

Asking that these rules be changed raised several safety concerns. Not only would the CSM with attached LM and the SIVB be traveling near each other toward the Moon, but it was feared that the powerful impact of the SIVB might hurl debris high above the Moon into the path of the CSM and LM. We asked MSFC to determine if sufficient propulsion would remain after translunar injection so that we could steer the stage and if there would be any problems sending commands to control its trajectory. Douglas Aircraft Company, the SIVB manufacturer, had studied such an application of the Surveyor translunar insertion stage when it was thought that the Surveyor spacecraft would carry seismometers to the Moon, so some of the homework had already been done.

MSFC came back quickly with an analysis that it could be accomplished; it was only too glad to have this opportunity to demonstrate its engineering prowess and the versatility of one of its babies. At the end of May 1969 MSFC made a presentation to me and Michael Yates, and at the end of June we presented our case to the Change Control Board, providing the analyses showing that the SIVB could easily be commanded to hit at a preselected point and that debris from the impact would not threaten the LM and CSM.[14] Approval was given to proceed with the SIVB modifications, to the delight of the passive seismometer team. We would have to wait until *Apollo 13*, scheduled for an

April 1970 launch, before all the changes could be made to the SIVB and its command software to achieve the controlled impact.

After *Apollo 12*, the "rump GLEP" and GLEP came into conflict with the conservative MSC engineers. Some of the sites on our list for the remaining eight missions would require maximum performance from all the Apollo components. I can recall a contentious meeting at MSC, shortly after *Apollo 12*'s return, when the subject of future landing sites was on the agenda. This was a meeting of MSC managers and engineers to which a few of us from headquarters and Bellcomm were invited. Bob Gilruth, MSC center director, was the senior manager present, but the meeting was run, as usual, by Chris Kraft, Gilruth's newly appointed deputy, and by Jim McDivitt, manager of the Apollo Spacecraft Program Office. Jim, a recently retired astronaut, was an excellent manager and ran a tight ship. Among other qualities, he was noted for his famous (or infamous, depending on your point of view) daily status reviews, held in a conference room lined with displays and charts and devoid of chairs: no nonsense, get the information out, assign actions, and get back to work! The only bow to comfort was a long table down the middle of the room where you could set your coffee cup while you took notes. Based on his positive response to Petrone's earlier letter, we considered Jim relatively neutral in our debates on how to accomplish the best science. His major concern was always crew safety; if safety was not compromised, he would usually support our requests.

After the near pinpoint landing of *Apollo 12*, some of the constraints on site selection described in chapter 5 were relaxed, in particular the requirement for multiple sites to accommodate possible launch aborts. Only one backup site had been designated for *Apollo 12*, about thirteen degrees farther west, which would have allowed for a one-day recycle if a problem had occurred before launch. The rump GLEP and GLEP went through a process similar to our earlier deliberations to select high priority sites for landings after a successful *Apollo 12*. This time we came up with a new set A including seventy-two sites. We then narrowed the list to a set B of twenty-one sites and finally recommended twelve that included Fra Mauro for *Apollo 13* and even more challenging sites for missions 14 through 19. (By now the number of landing missions had been reduced by one, but we were still planning on a total of nine landings.)

But back to the meeting. Equatorial sites had been agreed on for the first three landings as the safest and most easily accessible, although the *Apollo 13* site, Fra Mauro, would be a little more challenging since it was surrounded by rougher terrain. These initial sites were within the "Apollo zone of interest." All were close to the equator and were covered by the greatest number of high resolution Lunar Orbiter photographs. Many uncertainties still existed in predicting the performance of the total Apollo system, but Bellcomm had already completed an analysis of SIVB, LM, and CSM performance showing that a high percentage of the Moon's nearside could be reached while maintaining the required safety margins.

As the meeting droned on and such things as communication restrictions and propulsion budgets and margins were discussed, it became apparent that MSC management was going to take a conservative stand. Those of us who had been working on future landing sites were being asked (not quite directed) to rein in our expectations and continue to look for science sites near the Moon's equator. The nearer to the equator you landed, the more options were available to get you out of trouble. There was reluctance to go outside the "Apollo zone" despite the Bellcomm study. Besides, it was a Bellcomm analysis, not one done by MSC engineers.

MSC's position was certainly understandable. Every mission was risky, from liftoff to splashdown, and a difficult lunar landing site only added to the risk. No one wanted to be responsible for the decision to land at a site where a crew would be lost, for whatever reason. An accident, such as befell the crew of *Apollo 1*, could result in the cancellation of the remaining missions, an outcome that few in NASA would have cheered. For the staff at MSC each flight involved more personal worries than, perhaps, for someone in Washington or elsewhere in the scientific community; crew members were their neighbors and co-workers. If a crew didn't return they would be living with the grieving families.

By this time I had many close friends in the astronaut corps and fully appreciated the danger inherent in each mission. However, Noel Hinners and I felt obliged to speak up. The only rationale for continuing the missions was to carry out good science, and this could be done only if we were allowed to explore sites far from the equator, sites already identified as having the potential to resolve important questions. We went so far as to predict that, based on Lunar Orbiter photographs, safe LM landing sites could be found almost anywhere on the Moon. If any other proof was needed, look at *Surveyor 7*, which, with minimum ability to target the landing site, had managed to land in rough terrain on Tycho's rim without an astronaut making last-minute adjustments. How much easier it should be with a man at the controls. These remarks were met with skepticism and grumbling from around the table, but this position was gaining support from many others, including some of the astronauts.

Eventually, as others with more clout weighed in, MSC management reluctantly agreed to process sites away from the equator. Undoubtedly each mission that lifted off after *Apollo 13*'s near disaster increased their anxiety; the chances of a major problem were rising with each flight. No matter how carefully we prepared, one or more of the five million parts included in every launch vehicle and spacecraft could fail or malfunction at any point in a mission.

On March 6, 1970, the Apollo Site Selection Board met at KSC to select the landing site for *Apollo 14*. With *Apollo 13* scheduled to land in the western part of the "Apollo zone," this was the first meeting of the board since the meeting described above. We looked on it as a test to see if MSC management would be swayed by our arguments and allow *Apollo 14* to land outside the "zone." Tony Calio, who had replaced Bill Hess as chairman of GLEP, presented the results of the GLEP meeting of February 6 and 7. GLEP recommended a site called Littrow, at the southeastern edge of Mare Serenitatis, well north of the "Apollo zone," and the MSC in-house site evaluation team recommended the same site. After several presentations, including two by Lee Scherer and Noel Hinners, the board approved the Littrow landing site, and Jim McDivitt signed off in agreement.[15] We had overcome the last hurdle toward planning the scientific exploration of the Moon during Apollo.

A key science ally at MSC was Jack Sevier. Jack's personality was perfect for the difficult job he was assigned, acting as a mediator between the scientists and MSC's engineers. Easygoing, with a ready smile and quiet sense of humor, Jack had been an important contributor to the rump GLEP meetings starting in 1967, providing MSC's views on the constraints that could affect site selection. He was the branch chief of the Operations Analysis Branch and as such was the focal point for all the competing factors that could influence the outcome of our scientific activities. He would later lead the Lunar Surface Planning Team for the J missions, which developed the astronauts' lunar surface timelines and ultimately shaped the successful outcome of each EVA.

With the missions still being scheduled rather rapidly and changes in their scientific content occurring with each mission, some members of the scientific community continued to publicly criticize how Apollo science was progressing. Soon after the return of *Apollo 11*, Gene Shoemaker was quoted as being highly critical of the way NASA management treated science on the Apollo missions.

This view troubled me deeply at the time: we had been working hard to expand the science, and I knew he was aware of how much more productive the next missions would be. There is no question they could have been better, but we had made great progress since he had first become involved. His statement drew the ire of Homer Newell and Rocco Petrone. Harold Urey, perhaps egged on by Tommy Gold, who always seemed to delight in knocking NASA, also criticized the lack of scientific input into NASA decision making.

In a letter to Newell in March 1970, Urey said he agreed with Gold "that well known people who have been concerned about the moon for years are so systematically neglected by the management of NASA."[16] He was particularly irate at their exclusion from the selection of Apollo landing sites. In regard to site selection he wrote that "the people who vote are loaded with geologists of a very limited view of lunar science," and he made a few other scathing comments. By this time, after just two missions, Urey was seeing the writing on the wall. His well-publicized theories on the Moon's origin were being proved wrong, and I suppose Nobel laureates don't like to be proved wrong. At a later date he would acknowledge his errors and even make jokes about them.

Newell's staff was asked to respond to Urey's letter, but they sent an information copy to our office. Rocco Petrone, not taking kindly to this criticism, asked that we address one of Urey's comments dealing with site selection. In my memo for Petrone's signature, which we hoped would be included in Newell's formal response, I named the scientists and engineers present and voting at the last Site Selection Subcommittee meeting at MSC.[17] I listed twenty-one names: three geologists, two astronomers, four geophysicists, three NASA engineers, two geochemists, one nuclear chemist, three physicists, one geodesist, one atmospheric physicist, and one cosmologist-chemist, Harold Urey. Of the twenty-one, nine were government employees or contractors and the other twelve came from universities or private research laboratories. All had been involved in lunar research for at least the past five to ten years, which pretty well covered the period when interest in the Moon became widespread. Urey had picked the wrong topic—site selection—to complain about, but his overall concern had some merit. His complaints and those of others were primarily a criticism of how MSC was interacting with the scientific community, which once again was becoming contentious after Tony Calio replaced Bill Hess.

Urey's letter came just one month after a meeting at MSC when a group of scientists, all closely involved in Apollo investigations, met with MSC management to discuss the problems they were having working with MSC staff. Urey had not been invited to this meeting, nor had Tommy Gold, which may have added to their pique; Newell attended as an observer. After the meeting Newell apparently thought the situation was resolved and wrote Gilruth a complimentary letter; but he didn't really understand the depth of distrust that was building between Calio's organization and the scientists who were devoting more and more of their time to making each mission as successful as possible. Yet the meeting was useful in making McDivitt and Chris Kraft more aware of the needs of the scientists, and relationships with their offices improved. MSC agreed to arrange for more time in the astronauts' schedules so the PIs could explain their experiments and their requirements during deployment or operation. The PIs also asked for a better system of communication between the scientists in the SSR and the crews. They cited difficulties that arose during *Apollo 12*, when it took ten to fifteen minutes for questions raised in the SSR to be relayed to the astronauts by the CapCom, if they went out at all—and often they didn't.[18] There was some improvement on succeeding missions, but in general MSC and the Flight Operations Directorate (FOD) tended to ignore this latter request. Flight directors and CapComs felt, with some justification, that they shouldn't interrupt the crews on the surface with a lot of questions and directions; they had enough to think about.

Apollo 13 was a scientific disappointment but an engineering triumph. We lost a precious ALSEP (one of only six purchased), but the opportunity to study this site and collect valuable samples was realized when *Apollo 14* went back to the *Apollo 13* landing site. In spite of this disappointment, I never heard any complaints from the PIs, many of whom had worked with Jim Lovell, Fred Haise, and John "Jack" Swigert to prepare them for their flight. Like everyone else, we could only cheer the skill of all the NASA engineers and support contractors who brought the crew home safely. The *Apollo 13* crew members who performed so well under the threat of being the first astronauts to die somewhere in space, and the many heroes in the FOD led by Eugene Kranz, have had their roles well documented, so I will not try to add to that story. Science probably gained from the failed landing. It helped us refocus on how important each mission was. There were no givens; we had to make sure the remaining missions would be fruitful. And it seemed to make management more receptive to our requests to improve the science content of the last missions. The drama of *Apollo 13*'s rescue also ensured a more attentive public for the next missions and a wider audience interested in what we were discovering.

One experiment, the passive seismometer left behind at the *Apollo 12* landing site, did achieve important results from *Apollo 13*. Despite the problems the crew encountered during the rest of the mission, the *Apollo 13* SIVB stage, the first programmed to strike the Moon, accomplished its job by landing some eighty-five miles from the *Apollo 12* ALSEP. The seismometer received strong signals, and the impact had so much energy—estimated to be the equivalent of twelve tons of TNT (larger than the LM ascent stage impact because of its greater mass and higher velocity at impact)—that it sent seismic waves deep into the lunar crust. This elated Gary Latham, the passive seismometer PI, because he and his coinvestigators could now make some preliminary estimates about the Moon's deep structure.

When Lee Scherer's office was formed at the end of 1967, several of us involved in lunar science planning left Advanced Manned Missions, but Phil Culbertson stayed, eventually becoming director. In March 1970 he negotiated a memorandum of understanding with the Apollo Program Office to work cooperatively on lunar planning in case funding became available to continue missions beyond the scheduled Apollo flights.[19] Our two offices continued working jointly on post-Apollo planning for several more years, despite the lack of official sanctions to build the hardware needed for extended missions.

After *Apollo 13* failed to land, and reacting to the increasing clamor in some circles to halt the missions, in July 1970 our office issued a summary report of what we had learned to date from all our missions, manned and unmanned, and where we thought lunar exploration should be going.[20] The objective of the report was to support Culbertson's planning efforts in Advanced Manned Missions and to present an "Integrated Space Program Plan" that would provide mission schedules extending to 1990. It represented our last effort to justify a continuing program of manned and unmanned exploration by building on Apollo and other programs, including Mariner and Viking, and factoring in programs on the drawing boards such as Skylab and space stations. We presented an integrated program that included lunar bases and manned interplanetary launches.

Recently a quotation from Charles Lindbergh came to my attention. Asked about the $25,000 Orteig Prize offered for the first nonstop flight between New

York and Paris, which he won with his daring flight in 1927, he responded, "The important thing is to start: to lay a plan, and then follow it step by step, no matter how small or large each one by itself may seem." One could make a reasonable argument that Lindbergh's successful flight was the first step toward today's commonplace travel across the Atlantic and to almost every point on the globe. With Project Apollo we had taken the first step in mankind's leaving Earth and exploring our solar system. We believed we had put forth a step-by-step program to build on Apollo and move logically to the next objectives: space stations, lunar bases, and manned flights to Mars as early as 1989.

No such logical plan was ever agreed to. Some administrations have ignored space exploration, and some have paid it lip service. In the end, a program that would take advantage of the expertise and capabilities developed for Apollo was never endorsed. The report is now resting in one of my dilapidated packing boxes, perhaps the only surviving copy of our vision of a long-range plan for exploring the solar system. It was grandiose—undoubtedly too grandiose for the times—but in 1970 everything we proposed was achievable based on the technology in hand. All that was needed was the leadership to commit the nation to the next step.

In January 1971, just two weeks before the scheduled liftoff of *Apollo 14,* the second lunar science conference was held at the Lunar Science Institute. Although many of the same people attended this conference as were at the one held after the study of the *Apollo 11* samples, the sense of excitement was missing. The only new samples that had been studied, aside from a few grams of material brought back by the Soviets' *Luna 16,* were those returned by *Apollo 12* a year earlier. Whereas restrictions had been placed on the release of information about the *Apollo 11* samples, the *Apollo 12* sample PIs were not prevented from publishing the results of their studies of material returned by the mission. Most of the new information was already public and well known by the attendees.

The big debate at the conference dealt with the significance of the high content of radioactive elements (uranium, thorium, and potassium 40) found in some of the *Apollo 12* samples, which would imply an early, very "volcanic" Moon. There were other differences from the *Apollo 11* samples, suggesting that the Moon may have had an unusual differentiation history. It also appeared after initial study that the mare material sampled at the *Apollo 12* site was about a billion years younger than that collected at the *Apollo 11* landing site, suggesting that the Moon had gone through several major periods of mare formation. These findings would continue to be debated as each mission brought back new information.

Apollo 14 was sent to the site chosen for *Apollo 13*, Fra Mauro, in a hilly, upland area just a short distance east (112 miles) of the *Apollo 12* site. From the perspective of our plans to deploy the ALSEPs in a broad network so we could triangulate on phenomena at the Moon's surface or occurring at depth, being so near the *Apollo 12* ALSEP was not ideal. But from a geological point of view it was considered an important site, since we believed that the samples returned would include debris ejected from the huge Imbrium basin to the north. Again, as for *Apollo 12*, we hoped to collect samples from deep within the Moon that would help resolve some of the questions raised at the second lunar science conference. They would also be useful to Gary Latham and his coinvestigators in interpreting the Moon's deep structure, since these rocks would tell them how fast the seismic waves created by the SIVB impacts should travel compared with what they were observing in the records received back on Earth.

Had *Apollo 13* been successful, we were willing to accept the deployment of the ALSEP so close to *Apollo 12*. It was to be the last of the landings near the Moon's equator, reflecting MSC's cautious approach. We had expected that after *Apollo 13*, *Apollo 14* would land at Littrow, the first site selected solely for its scientific value and, because it was far off the lunar equator, ideal for our ALSEP network. The geological rationale for landing at Fra Mauro still held, but the decision to retarget *Apollo 14* there was doubly painful from a scientific perspective. With the loss of *Apollo 13*, there were only six more projected landings (ultimately reduced to three) to uncover the Moon's secrets hidden on or below a surface area roughly equivalent to all of North and South America combined. And well over half of that area was inaccessible because it was outside our landing capabilities or on the Moon's farside. Imagine trying to understand those two continents with only six widely scattered small points of knowledge!

The landing site was to be within walking distance of what appeared to be a crater of recent vintage, named Cone by the Field Geology Team because of its steep, funnel-like inner slopes. From the Lunar Orbiter photos we could see large blocks on the rim of Cone Crater, reinforcing the belief that if the astronauts could get to the rim they would be able to sample Imbrium ejecta in the rocks "mined" by the Cone Crater impact. Alan Shepard guided the LM to a perfect landing within two hundred feet of the target point and less than a mile from Cone Crater, whose rim could be seen in the distance when he and Edgar Mitchell descended the LM's ladder. This time there was a color television camera, with better resolution than the *Apollo 11* camera, and it functioned well, providing views of the astronauts as they climbed down to the surface and panoramas of the landing site as they worked near the LM.

Between the *Apollo 13* and *Apollo 14* launches we had built a small two-wheeled cart, the modularized equipment transporter (MET) discussed in chapter 8, to help the astronauts carry all the gear that was now part of the field geology experiment. It was unloaded from the LM descent stage near the beginning of the first EVA, and the crew stowed the tools and equipment they would need for the sampling scheduled on the first EVA and the traverse to the rim of Cone Crater, the major objective of the second EVA.

The first EVA went off with no big hitches, and the major tasks—the ALSEP deployment and sample collection near the landing site—were successfully completed. A new experiment, the active seismic experiment, was conducted in conjunction with the ALSEP deployment. Three geophones were strung out on cables to the south of the ALSEP, with the last one approximately three hundred feet from the ALSEP central station. The first part of the experiment consisted of setting off small charges, about the size of a shotgun shell, housed in a hand-held "thumper" hardwired to the ALSEP central station electronics, which provided timing data and transmitted the signals received by the geophones back to Earth. Mitchell carried the thumper out to the last geophone and set off a charge, then retraced his steps back to the geophone closest to the central station, setting off charges along the way. Twenty-one charges were scheduled, but a few misfired and only thirteen were recorded. A second part of the experiment consisted of a mortar designed to fire four small explosive charges various distances away from the geophones, the farthest to land five thousand feet from the mortar. This second part of the experiment was not conducted until many months later, to avoid any possibility that the mortar fire might damage the nearby ALSEP central station.

Although Shepard and Mitchell could see the ridge formed by Cone Crater in the distance when they started out on the second EVA, once they began walking and pulling the MET they soon lost sight of the ridge behind the intervening low hills and hummocks. Others have described in some detail their difficulties in reaching the rim of Cone Crater. They didn't quite make it, but they came close, and they sampled ejecta thrown out by the impact that formed the crater, the main geological objective of the mission. After the difficulties they encountered attempting to reach Cone Crater's rim, they probably both wished they had the LRV that would be carried on the next mission. Another new experiment on this mission, the Lunar Portable Magnetometer, was operated twice during this EVA, and the readings were relayed back to Houston by voice. The samples collected during both EVAs weighed almost ninety-five pounds.

Like the CM pilots before him, Stuart Roosa carried out several experiments on the way to the Moon and while the other astronauts were on the lunar surface. The number of experiments assigned to the CM pilot was increasing with each mission as we attempted to take full advantage of his time and the added payload weight that was becoming available. Roosa completed several new photographic tasks and other types of experiments. Bellcommers Farouk El Baz and Jim Head took on growing roles instructing Roosa, as well as the CM pilots on the final three flights, in the objectives of the photographic experiments and the cameras' operation. After the film was returned, they also helped interpret the data obtained. *Apollo 14* marked the end of the H missions, one short of the four originally planned.

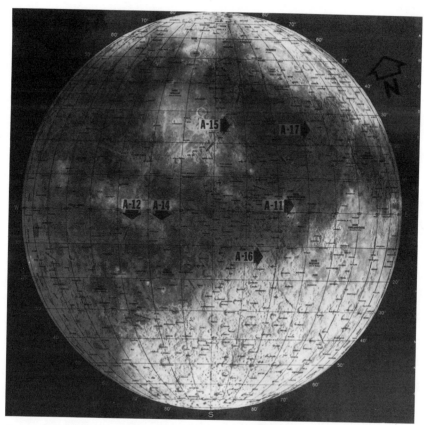

Composite full Moon photograph showing Apollo landing sites. (NASA S79-27140)

Astronaut Edwin E. Aldrin stands near the solar-powered passive seismic experiment that was deployed on *Apollo 11*. (NASA AS11-40-5948)

The Laser Ranging Retro-Reflector (LRRR) deployed on *Apollo 11*. The lunar surface close-up camera stands to the right of the LRRR. (NASA AS11-40-5952)

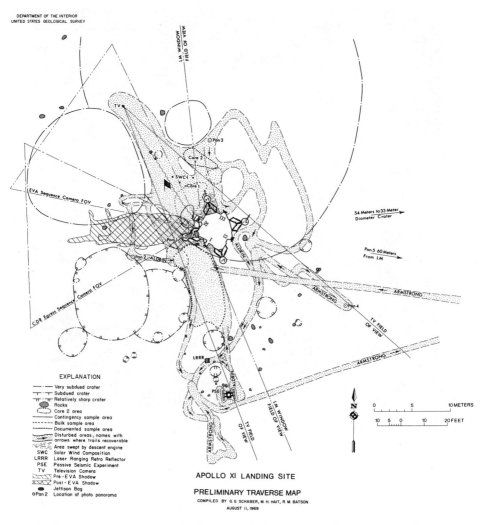

LM WINDOW FIELD OF VIEW

TV

Pan 3

Core 2

SWC
Core 1

EVA Sequence Camera FOV

54 Meters to 33 Meter Diameter Crater

ARMSTRONG

Pan 2 ALDRIN

ALDRIN

Pan 5 60 Meters From LM

Pan 1

Pan 4

ARMSTRONG

ARMSTRONG

CDR Egress Sequence Camera FOV

TV FIELD OF VIEW

LRRR

ARMSTRONG

NORTH

PSE

LM WINDOW FIELD OF VIEW

TV FIELD OF VIEW

N

EXPLANATION

- Very subdued crater
- Subdued crater
- Relatively sharp crater
- Rocks
- Core 2 area
- Contingency sample area
- Bulk sample area
- Documented sample area
- Disturbed areas; names with arrows where trails recoverable
- Area swept by descent engine
SWC   Solar Wind Composition
LRRR   Laser Ranging Retro Reflector
PSE   Passive Seismic Experiment
TV   Television Camera
- Pre-EVA Shadow
- Post-EVA Shadow
- Jettison Bag
Pan 2 Location of photo panorama

0   5   10 METERS

10 5 0   10   20 FEET

APOLLO XI LANDING SITE

PRELIMINARY TRAVERSE MAP

COMPILED BY G. G. SCHABER, M. H. HAIT, R. M. BATSON
AUGUST 11, 1969

Preliminary traverse map of the *Apollo 11* landing site prepared by the Field Geology Team.

The Apollo Lunar Surface Experiments Package (ALSEP) central station deployed for the first time by the *Apollo 12* crew. The antenna points toward Earth so that the central station can receive commands and control all the experiments. The passive seismic experiment is to the left of the station, the lunar surface magnetometer in the background. (NASA AS12-6428)

APOLLO 15 FLIGHT PROFILE

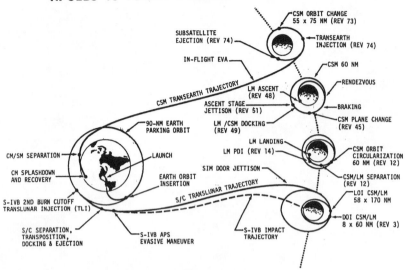

The *Apollo 15* flight profile is typical for all landing missions. (NASA M-933-71-15)

The Mission Operations orbital Science Support Room during the *Apollo 15* mission. On the left, leaning on the console, Isadore Adler, principal investigator for the X-ray fluorescence experiment, talks to NASA's George Esenwein and David Winterhalter. (NASA S71-43255)

The *Apollo 16* ALSEP with the lunar surface magnetometer in the foreground. In the background, astronaut John W. Young prepares the active seismic experiment. (NASA AS16-113-18373)

The active seismic experiment mortar box holds four mortar shells. Below the mortar box is the orientation template. (NASA AS16-113-18380)

An astronaut uses a lunar rake at the *Apollo 16* landing site to collect small rocks from the soil. (NASA AS16-116-18690)

The Field Geology Team monitors the *Apollo 16* mission from the Mission Operations surface Science Support Room. *Right to left,* Dale Jackson, James Head (standing), principal investigator William Muehlberger, two interested spectators standing in the corner, *Apollo 17* crew Eugene Cernan and Harrison H. Schmitt, and other staff. (NASA S72-37410)

The far UV camera/ spectrograph, the first astronomical instrument placed on the lunar surface, stands in the shadow of the *Apollo 16* lunar module. (NASA AS16-114-18439)

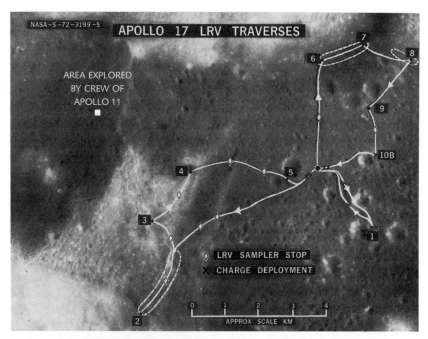

APOLLO 17 LRV TRAVERSES

AREA EXPLORED
BY CREW OF
APOLLO 11

7
6
8
9
10B
4
5
3
1
2

◇ LRV SAMPLER STOP
✕ CHARGE DEPLOYMENT

0   1   2   3   4
APPROX SCALE KM

Apollo 17 planned traverses at Taurus-Littrow landing site. Sampler stop and charge deployment symbols indicate sampling stations and locations for the Seismic Profiling experiment explosive packages that would be detonated after the crew departed. Small box in upper left compares, at same scale, the area sampled by the crew of *Apollo 11* at Tranquillity Base. (NASA-S-72-3199-S)

In this close-up of the gnomon at the *Apollo 17* sampling site, astronaut Harrison H. Schmitt uses an extension handle attached to a scoop, just out of sight behind a rock. (NASA AS17-145-22157)

A Lunar Receiving Laboratory's sample-handling cabinet undergoes a leak check. For *Apollo 11*, the cabinets operated under a vacuum to avoid the escape of lunar material, and the gloves would have been forced inside. To simplify processing, material returned after *Apollo 11* was examined in cabinets containing nitrogen at one atmosphere. (NASA S64-33455)

The "genesis rock." Returned by *Apollo 15,* it is one of the oldest samples from the Moon, giving an age of crystallization of approximately 4 billion years. (NASA S-71-42951)

The J Missions:
We Almost Achieve Our Early Dreams

Apollo 15 was the first of the J missions. Years of struggle and cajoling, as well as long hours spent meeting with contractors, principal investigators, scientific committees, and NASA colleagues, had finally borne fruit. All the allowances for payload, extravehicular activity time, distance traversed, and sample return would suddenly double or triple. Although we had greatly increased our ability to explore, however, there would be no dual launches, no two-week exploration timelines to construct, and no seven-thousand-pound science and logistics payloads that would have given us the experience to plan for lunar bases. Our attempts to convince Congress and the Nixon administration to extend lunar exploration had failed. Instead of the five more missions we had been planning just six months earlier, only the three J missions remained. After *Apollo 17* returned, Project Apollo would close its doors. We chose to put this sad ending out of our minds and concentrate on ensuring the success of the last missions.

Two days before the launch of *Apollo 15* several of my colleagues and I flew to Orlando and then drove to Cocoa Beach, Florida, to prepare for the prelaunch press briefing. George Esenwein and Floyd Roberson would describe the new command and service module experiments; Ben Milwitsky and Richard Diller would do the same for the lunar roving vehicle; and I would cover the surface science. The routine at these briefings was that we would make short prepared statements, illustrated with vugraphs, and then take questions. Gene Simmons joined us, having recently transferred from MIT to the Manned Spacecraft Center, and some of the PIs—those with experiments flying for the first time— were on hand to discuss them. Gene, with his new title of chief scientist and an assignment to once again try to improve relations with the scientific community, had written a guidebook, "On the Moon with *Apollo 15.*" It was sought after by the media as a quick reference covering the science aspects of the mission and providing some easy quotations, a service the hardworking members of the press always appreciated. Gene compiled similar guidebooks for the last two missions, *Apollo 16* and *Apollo 17.*

NASA's Public Affairs Office usually released a mission press kit about ten days before the launch to give the media a chance to get familiar with the mission. It included details on all aspects, including the mission's scientific activities and experiments. Before these last briefings at Cocoa Beach, media briefings for each launch were conducted at intervals at places such as MSC and Kennedy Space Center, and a major briefing was always held at headquarters about a month before launch for the large Washington media contingent. But the various briefings at Cocoa Beach (other parts of the flight besides the science were covered) the day before the launch were always the best attended, resulting in a lot of print and sound-bite coverage. Some seventy-five members of the media attended our briefing, held in a large second-floor conference room at the Friendship Inn the morning of July 25. Jack Hanley and Don Senich came along to answer any questions on two new pieces of equipment, the lunar drill and the soil mechanics penetrometer.

By this time our office staff had grown, mostly with members detailed from other agencies or NASA centers. NASA budgets had been going down for the past four years, and with those reductions came a semifreeze on hiring NASA civil servants. We always argued, to no avail, that some aspects of NASA business were still growing—Apollo science as an example—and needed more bodies. To spread the added workload in our office, we obtained detailees from the Army Corps of Engineers, the United States Geological Survey, and the Jet Propulsion Laboratory. In addition to Jack Hanley from USGS and Don Senich from the Corps of Engineers, USGS lent us Gerald "Jerry" Goldberg, and JPL sent Ewald Herr and later Peter Mason and Ronald Toms for assignments that lasted one year or more. Hanley, Goldberg, and Senich stayed with our office for over three years and were invaluable additions, cheerfully (usually) taking on the "dog work" that every government bureaucracy generates as well as the more interesting oversight for science payload development.

Besides the excitement of the upcoming launch, which brought media representatives from all over the world, in the days before the launch Cocoa Beach and the surrounding area were the site of many parties, a tradition that went back to the first rocket launches in the late fifties. These parties grew with each new manned launch. By custom, the night before the launch the big companies, with a major stake in the mission, would hold open houses that included food and drink. For Apollo launches, North American, Douglas Aircraft, Grumman, and Boeing, as well as smaller companies such as Bendix (many of these companies have since merged and lost their identities), would all hold their own affairs to tout their participation, with some competition to throw the best party. Similar "splashdown" parties were held in Houston, near MSC, after the end of each mission, and these would be even wilder, if that was possible, than the prelaunch parties at the Cape. These MSC parties were by invitation only, and invitations were always in great demand.

Up and down the beach a half dozen or more parties would go on into the wee hours. The morning after would be spent describing some of the more outrageous events. Nothing attracts the press more than a free party and, I might add, the many VIPs and sightseers who were in attendance. We civil servants tried to be discreet, but we would also drop in on a few of the parties even though some of us had official duties the next morning. With the *Apollo 15* launch scheduled for 8:34 A.M., it meant waking up early to beat the traffic and get to my VOA broadcast site. I had been promoted since *Apollo 11* and was now a full partner in the broadcasts.

The TV networks were getting more and more elaborate with their coverage of each succeeding mission. For *Apollo 15,* with its promise of real-time TV pictures during the astronauts' LRV traverses, the major networks had assembled simulated lunar terrain to illustrate how the astronauts were going about their exploration. NBC had a small working model of the LRV, and CBS and ABC had borrowed full-scale working models. For TV and news media pools not fortunate enough to have their own models, we supplied static mock-ups of the LRV and Apollo Lunar Surface Experiments Package at MSC, where their reporters could be shown standing in front of the models.[1] Each of the major networks also had a captive astronaut in the studio to explain the intimate details of what was going on during the mission.

In terms of science training, the crew of *Apollo 15* was the best prepared yet. Based on my observations, Dave Scott, James Irwin, and Alfred Worden showed the greatest interest of any of the crews to date in understanding the science objectives for their landing site—the Apennine Mountains and nearby Hadley Rille—and the new suite of experiments housed in the lunar module and CSM.

Scott, as mission commander, set an example for his two crewmates through his hard work and contagious enthusiasm. Hadley Rille, a long, sinuous valley, was one of the most intriguing features on the Moon's surface. It was surpassed in interest only by the first landing site, when any information returned was bound to be extraordinary. Several theories had been proposed to explain the rill's origin: that it had been formed by water discharged from the Moon's interior (the least favorite theory among most lunar scholars); that it was a lava channel or collapsed lava tube or had been gouged from the surface by some volcanic event; or that it was the remnant of faulting or stretching of the lunar crust.

The landing site was on the eastern rim of the Imbrium basin, so we anticipated that the returned samples would include material from the Apennine Mountains, probably formed by uplift and ejecta from deep in the Moon's interior. Other samples should include Imbrium basin fill consisting of some type of lava. Samples from the edge of Hadley Rille might resolve the question of its origin. We might even collect samples of the ejecta from Mare Serenitatis, just a short distance to the east. If we could identify their source, age dating these samples would go a long way toward explaining key events in the Moon's early history that shaped its final form.

Adding to our excitement about this mission was the greatly increased radius of operation for the astronauts and the many new experiments that would be performed. The science payload delivered to the lunar surface would be almost 1,200 pounds, compared with the *Apollo 11* payload of less than 200 pounds and more recently the 470 pounds carried on *Apollo 14.* If the mission went as planned, *Apollo 15* would come closest to our earlier dreams for the first post-Apollo missions. *Apollo 15* marked another milestone: Jack Schmitt was named to the backup crew. Based on previous crew rotations, this would have put him in position to be named to the prime crew for *Apollo 18,* now canceled, but at least he had moved up in the pecking order. This would be Gordon Swann's last mission as PI for the field geology experiment. He and his many Flagstaff colleagues and coinvestigators, which included Bill Muehlberger, the designated PI for the last two missions, had been building to this climax after many years of hard work. Next best to being on the Moon themselves, the J missions would validate their efforts, and they worked tirelessly to prepare the crew.

All the preparation paid off. The crew performed flawlessly. The Science Support Room (SSR) geology team, led by Swann, listened, recorded, debated, and attempted to interpret in real time everything that was happening on each of the three EVAs during the 67 hours the crew was on the surface. Total time of the three EVAs was 18.5 hours, a new record. As we had practiced at Flagstaff, Scott also performed the first and only stand-up EVA when, shortly after landing, he opened *Falcon*'s overhead hatch and stood on top of the ascent stage engine cover to get a bird's-eye view of the landing site. While enjoying the scene around him, he took some panoramic pictures and planned the upcoming traverses. With only minor upgrades of LM systems for the J missions, we were able to more than triple the time the *Apollo 11* crew had spent on the lunar surface. The *Apollo 15* EVA time was almost as long as the total time the *Apollo 11* astronauts spent on the Moon. These numbers confirmed in my mind that the one- to two-week visits we envisioned for the post-Apollo dual-launch missions could have been achieved by making the modifications to the Apollo systems we had studied.

Another change for *Apollo 15* was the inclusion of scientist-astronauts Jack Schmitt, Joe Allen, Bob Parker, and Karl Henize as capsule communicators; Allen, who had served as mission scientist, usually manned the console during the EVAs. On all the previous missions only Schmitt, for *Apollo 11,* had been given this high profile task. Although it is difficult to point to any specific advantages of having them at the consoles, interaction between the ground and the crew of *Apollo 15* was lively, and certainly we in the SSR felt more comfortable knowing that Allen could immediately interact with the crew if necessary. We passed suggestions and questions to the CapComs, and in contrast to *Apollo 12,* many were passed on.

The scientific harvest from *Apollo 15* was spectacular, derived both from the lunar surface and from lunar orbit. An ALSEP was deployed at the northernmost point reached by any Apollo mission. This ensured an ideal positioning of the ALSEPs for triangulating readings for experiments like the passive seismometer and LRRR that needed site separation and the surface magnetometer that was attempting to discover if the Moon's magnetic field might vary from site to site. The three LRV traverses covered almost seventeen miles, during which the astronauts studied twelve locations in addition to the immediate landing site. They collected almost 170 pounds of samples and took more than 1,100 photographs. Besides the many photographs taken with the Hasselblad cameras, we also had TV coverage of eight stations and some footage taken while the astronauts were under way on the LRV. Eight major experiments were conducted in lunar orbit, and many types of photographs were taken from both the CM and the scientific instrumentation module (SIM) bay in the service module, adding to the wealth of new information that included coverage of the Moon's farside.

Apollo 15 recorded one other first, and last. After Scott and Irwin's rendezvous with Worden back in lunar orbit, Lee Silver was called out of the SSR to go to the Mission Operations Control Room. Lee had led many of the field training exercises for the *Apollo 15* crew and had established a close rapport with them. Scott wanted to talk directly to Lee to thank him for his long hours and dedication to their education. Thus, for the first and last time during an Apollo mission, a member of one of the science teams talked directly to the astronauts while a mission was in progress without going through an astronaut CapCom. Lee passed on our congratulations and told them how excited we were about what they had accomplished, and he reported the results we were already seeing as we reduced the traverse data. I hope that when we return to the Moon such exchanges between scientists on Earth and those working on the lunar surface will be the norm, for it will surely add to the value and efficiency of future lunar exploration.

All the experiments and equipment carried on *Apollo 15* performed up to expectations except for the drill. To quote from the crew's observations, "The deep core could not be extracted from the uncooperative soil by normal methods; the two of us, working at the limit of our combined strength, were ultimately required to remove it." The exterior flutes contributed to this condition because the drill stem was pulled into the ground still deeper when the motor was activated.[2] This activation was supposed to clear the flutes for easy extraction.

As we were to discover, there were two complications involving the drill. The first and most serious occurred as the astronauts were drilling the bore holes for the heat flow experiment, and they encountered the second while trying to extract the core sample. Mark Langseth, the heat flow PI, had designed his experiment around placing the sensors in a cased drill hole some ten feet deep. When Scott attempted to drill the first hole he could not go much deeper than about five feet, well short of his target. The drill stem refused to go any farther no matter how hard he tried. Frustrated and thinking he might have hit a large rock, he stopped working on the first hole and tried to drill the second. Again, he could not get penetration much below the first length of drill stem. Time was fleeing, so he was instructed to stop drilling, place the sensors in the holes he had, and finish the remaining first EVA tasks. Langseth ended up with his sensors much closer to the lunar surface than he wanted, and he feared he would not get the high quality data he was hoping for. More later on the tribulations of the heat flow experiment.

In drilling the core sample the astronauts encountered a different problem. The drill penetrated to the full depth quite easily—too easily, it turned out. Fortunately, through the crew's "combined strength" they salvaged the core. Thus the time had not been wasted. Wasting time during a mission was not attractive to anyone, especially when the medical team monitoring the astronauts could see they were in danger of exceeding their physical limits while trying to remove the core. As the astronauts struggled, recommendations were made in Mission Control to abandon the attempt, but Scott and Irwin persisted and saved the day.

Immediately after this well-publicized glitch, while the crew was still on the Moon, Rocco Petrone caught me in the Mission Control Center and issued one of his famous edicts—we were to solve the drill problem before the *Apollo 16* flight readiness review! I huddled with Jack Hanley to discuss our course of action. Following our usual method of addressing such issues, we appointed a "tiger team," consisting of Dave Carrier and several other MSC engineers plus Jack Hanley and Don Senich from my office. They were dispatched to Denver to meet with Martin Marietta, find out what went wrong, and make the necessary modifications. The solution had to be found quickly, since the drill had to meet the *Apollo 16* equipment stowage window for its preflight checks. We also had to be prepared to modify the astronauts' training and simulation schedules if any changes required new instructions or training.

After discussing the crew's observations and reviewing how the drill had performed, we knew we had two separate problems, one in drilling the bore holes for the heat flow experiment and the second in extracting the deep core. The tubular drill sections used for the heat flow holes were of a different design than the core stems. Since the core stems had drilled to almost eight feet without any trouble, this design difference had to hold the clue. Langseth thought he knew what had gone wrong. The flutes on the outside of the heat flow drill sections did not extend the full length of each section; they stopped short of the ends, leaving an open space on the shafts.

With new information on the characteristics of the lunar soil derived from the soil mechanics experiment, the fidelity of the simulated lunar soils used during testing was improved, and we conducted many tests, some in a vacuum chamber. Langseth was right. The short interruption of the flutes at the joints had prevented the cuttings from traveling up the tubular sections; they jammed at the joint after the next section was added and the joint was drilled a short distance below the surface. The design had worked satisfactorily during our terrestrial trials, but the soil simulant used during the original tests had not been as compact and dense as real lunar soil.

The original design of the heat flow tubing had called for titanium inserts at the ends of the fiberglass sections to strengthen the joints so they could be screwed together like the core sections. This would also ensure good meshing of the flutes, but Langseth had been concerned that so much metal in the tubing might disturb the sensitive readings he was hoping to obtain. We had removed the titanium joints, and the tubing carried on *Apollo 15* required that each section be pushed into the next. The flute alignment was determined by how carefully the astronauts joined the sections, but there was always a small gap between the flutes. Langseth agreed that we would have to go back to the original design for *Apollo 16*. This was done, and the drilling tests were successful.

We believed we had solved the heat flow drilling problem, but we needed more tests to understand the core extraction problem. During the crew debriefing it emerged that because Scott had had trouble drilling the heat flow bore holes, he had perhaps put too much pressure on the drill while coring. This forced the core stem into the soil before the flutes could completely clear the cuttings, thus jamming the core stem. Tests at Martin Marietta showed that if the drill penetrated more slowly, even in a more moonlike soil than we had used in previous simulations, there should be no difficulty extracting the core on the next missions. Just in case it did jam, we designed a new device to jack the drill stem out of the hole if it would not come free using the normal procedure of rotating the drill core in place. Score two hits for the tiger team and the Martin Marietta and Black and Decker engineers. On *Apollo 16* and *Apollo 17* the drill worked well for both applications.

Although the missions, starting with *Apollo 15,* were being launched on a more relaxed schedule, approximately one every eight months, there was a heavy training burden on everyone at MSC, KSC, and the Field Geology Team. At headquarters we also felt the pinch as we tried to stay abreast of the progress, or lack thereof, so we could keep Petrone and other senior management up to date. Rocco hated surprises, and this attitude carried over to all his staff and was often reflected during his weekly status reviews. These reviews, held at our offices at L'Enfant Plaza in a large, windowless room lined on both sides with multiple sliding status boards, would consume half a day or longer depending on the number of outstanding issues. The status boards were updated daily through a contract with the Boeing Company and were used extensively during the reviews. Each Apollo office would make a presentation so that Rocco could get a snapshot of the program covering everything from spacecraft and payload status to funding, manpower, and eventually, final plans for close-out. After the near disaster of *Apollo 13* and with the last missions firmly scheduled, the atmosphere was getting tenser; we had to make sure that nothing was left to chance and that no dumb mistake would jeopardize a crew.

In November 1971, four months after the return of *Apollo 15,* Lee Scherer was transferred to a more prestigious management position, director of the NASA Dryden Flight Center at Edwards Air Force Base in California. Don Wise, Lee's deputy, tiring of the Washington scene, had already gone back to academia. O. B. O'Bryant was named to replace Lee for the final two missions, and I was named program manager for Apollo surface experiments, including the ALSEP, since Ed Davin and Dick Green had left to take new jobs at the National Science Foundation.

The crew of *Apollo 16*—John Young, Charley Duke, and Kenneth Mattingly—had been named long before *Apollo 15* was launched, and their training and simulations overlapped those of the *Apollo 15* crew. For the Field Geology Team there was the added consideration of changing the PI from Gordon Swann to Bill Muehlberger, with Bill adding some new coinvestigators from USGS and Bellcomm. This transition went off without a hitch, since both Gordon and Bill had worked together for a long time and there were no professional jealousies involved in the switch. With the overlap in training the *Apollo 15* and *Apollo 16* crews, it was difficult even to notice a change, and almost all the faces remained the same. With each mission, the training was also becoming more complicated, reflecting the added complexities of the J missions with their new experiments, both surface and orbital, and longer surface EVAs.

The *Apollo 16* crew had a much different character than the crew of *Apollo 15,* reflecting the personality of John Young. He had already flown three space missions, including the highly successful *Apollo 10,* and had more time in space than any other astronaut except Jim Lovell. Young was more relaxed than the hard-driving Scott. He was quick with a quip or story to break the intensity of a training or simulation session. In spite of the more relaxed atmosphere, the crew was required to spend hundreds of hours learning the scientific nuances of their chosen landing site, Descartes, a highlands crater near the Moon's center several hundred miles southwest of the *Apollo 11* landing site. In addition to learning what they should expect and look for at Descartes, they had to train to deploy the ALSEP and all the other experiments, including the redesigned drill and one new experiment, the far UV camera-spectrograph, our first chance to use the Moon for astronomical studies.

Bill Muehlberger, perhaps tempered by dealing with undergraduates during his tenure at the University of Texas, Austin, plus his long association with some of the eccentric personalities at USGS, meshed well with the crew. His team of coinvestigators and field geology instructors took on the task of instructing the crew by taking them to several training sites in the United States, Mexico, and Canada. The sites were chosen to expose them to field conditions representative of the latest geologic interpretations of what they might encounter at Descartes. Since this would be the first landing in the Moon's highlands, we expected to pin down their composition, an important determination in understanding the Moon's history. The crew would also sample the material that filled the large crater in which they would land; perhaps it was of a different composition than the maria sampled on the earlier missions. Some photogeologists studying the area around the landing site believed they were observing volcanic features, low hills that might be composed of lava or cinders, much like the formations just east of Flagstaff. Other interpretations were possible, but these hills looked unusual and, most important, were in the highlands. Young and Duke were conscientious students and quick learners. Those of us who tagged along to observe the training sessions were impressed at how well they were absorbing the huge amount of information thrown at them. The sessions included a trip to Sudbury, Canada (which I didn't attend), considered an excellent example of some of the geological situations they might encounter on the Moon.

Ken Mattingly, the CM pilot, who had been scratched from the *Apollo 13* crew because of fears he might have been exposed to German measles, for which he had no immunity, was the most studious of the three crewmen. He realized how fortunate he was to have a second chance, and he was determined to get the most out of the experiments he would operate from the CSM. Like Al Worden, he would be photographing and making measurements over a wide swath of the lunar surface. He spent many hours with Farouk El Baz, Goddard Space Flight Center PIs Isadore "Izzy" Adler and Jack Trombka, and other PIs, learning as much as he could about their experiments and what they hoped to achieve. The results of the *Apollo 15* orbital science were now available. Drawing on Worden's experience of operating the experiments for the first time, Mattingly was in a better position to manage them efficiently.

Apollo 16 was launched from KSC just before 1:00 P.M. on April 16, 1972, a more civilized hour for those of us covering the launch. There were no complications with the flight until after lunar orbit was achieved, but a major problem surfaced just after the LM and CSM separated. When Mattingly went through his checklist before firing the service module engine to circularize his orbit—his first order of business after separation—one of the gimbal motors that controlled the SM engine nozzle did not respond properly. If he could not get the gimbal motor to work, he would not be permitted to start the engine. Mission rules dictated that the landing would have to be aborted and the LM would rendezvous with the CSM using the LM ascent or descent engine or both. Once joined, LM propulsion would be used to get them out of lunar orbit and on the way back to Earth.

While Mission Control tried to find a solution, the LM and CSM were directed to orbit the Moon near each other but not to join up. The crew, and all of us sitting on the edge of our seats in the SSR, kept hoping the mission would not have to be aborted, but with every orbit that passed without instructions on how to proceed, it was becoming less and less likely that the landing would happen. If the landing was permitted, time lost would have to be deducted from the lunar surface staytime. Eventually, if too much time elapsed, the landing would have to be called off because the sun angle would impair visibility on the surface so Young would be unable to avoid small obstacles at the landing site.

After reviewing the data sent back to Earth from the CSM and consulting with the North American engineers, Mission Control decided it was safe to proceed. Mattingly successfully fired the engine to put him into the desired circular orbit, and Young and Duke completed their landing. The landing delay (approximately 5.75 hours) did reduce the time spent on the lunar surface. To make up for this lost time it was proposed to cancel the third EVA. After much pleading from Muehlberger's team, emphasizing the importance of the sampling sites selected for the third EVA, it proceeded as scheduled but was reduced by about two hours. To keep on the overall flight schedule, the time would be made up by lifting off from the Moon sooner after the astronauts returned to the LM at the end of the third EVA than originally planned.

With their problems behind them (there were a few others), Young and Duke went about exploring their landing site and deploying all their experiments. The ALSEP was set up during the first EVA, and this time the drill worked as designed. But the heat flow experiment met with calamity. After the first hole was drilled and the probe was lowered into the tubular casing, Young, working on another part of the ALSEP deployment, tripped over the cable connecting the probe with the central station and pulled it loose. ALSEP cables consisted of copper wires embedded in a thin plastic covering several inches wide, designed to lie flat on the lunar surface after they were unreeled from their containers. But after being coiled for several weeks in stowage, they tended to develop slight kinks. Whether that caused Young to catch the cable with his boot or whether he just misstepped, the cable came loose. After examining the end of the cable that had been torn off and describing it to Mission Control, it was decided it could not be reconnected.

We never anticipated a failure of this type, so no tools were carried for making a repair. Just to be sure we weren't overlooking a possible fix, and with a distraught PI begging us to find a solution, we put together another tiger team. But we could not come up with a guaranteed way to reattach the cable. With the cable broken the experiment could not operate, so we canceled the drilling of the second hole. This was a major setback for Langseth; the first deployment of his experiment had placed the probes too shallowly, compromising the readings, and this deployment was a complete failure. His experiment, considered one of the most important we would place on the Moon, would have one last chance on *Apollo 17*. But with the loss of this data point and the compromised data from *Apollo 15*, even if the *Apollo 17* deployment was successful our overall understanding of the heat flow from the Moon's interior would be open to question, since the measurements at the *Apollo 17* site might not be typical for the Moon as a whole.

The deep core was drilled successfully on the first EVA, and eight feet of core were recovered. The active seismic experiment, a duplicate of the one deployed during *Apollo 14*, functioned much better this time. We had made some minor changes in the thumper firing mechanism, and all twenty-one charges fired. The second part of the experiment, the mortar package, was placed in a better position relative to the ALSEP, and we were able to fire three of the four mortars one month after the astronauts returned home. After firing the third mortar, the pitch-angle sensor showed that the mortar box might have tilted, so we decided to hold off firing the fourth mortar until later.[3]

The far UV camera-spectrograph, the first telescope to be used on the Moon, was placed in the shadow of the LM and moved several times on succeeding EVAs to keep it in the shadow. It was pointed at different sectors of the sky, three times during the first EVA, four times during the second, and three times during the third, then the film was unloaded and returned to Earth. After setting up the ALSEP and other experiments during the first EVA, the astronauts returned to the LM and unloaded the LRV. This left just a short time for the first traverse, and they went to study and sample some craters less than a mile to the west.

On the second EVA the astronauts traveled south about two miles with a major objective of sampling the debris thrown out by a "recent" impact crater (named South Ray) about half a mile in diameter that had scattered ejecta a great distance in all directions. It was expected that this ejecta would provide us with good samples of a geological formation (named Cayley by USGS) that forms extensive highland plains thought by some to be volcanic in origin. If it proved to be volcanic, its composition and age would be important pieces of information for understanding the development of the lunar highlands that make up approximately four-fifths of the Moon's surface. Equal in importance to resolving the composition of the Cayley formation was obtaining samples of the mountain-making material in the vicinity of the Descartes Mountains. Although the mountains were some distance from the landing site, we believed we stood a good chance of recovering rocks deposited in the plains from impacts that occurred in the highlands.

Tony England, selected in the second scientist-astronaut class, was the mission scientist and CapCom during the EVA periods, and he did a good job of communicating with the crew and relaying questions and suggestions from the SSR. In his role as mission scientist, he had accompanied the crew on many of their training trips and participated in the simulations leading up to the mission, as Joe Allen had for *Apollo 15*. He was intimately familiar with all the equipment and experiments and was able to quickly give advice when needed.

SSR operations had improved with each flight. Beginning with *Apollo 15*, we could supply more and more backup information. We kept careful track of the EVAs as they progressed; the planned traverses with each station were identified on a three-dimensional model constructed from the Lunar Orbiter photographs, and we had a list of planned activities for each stop. We also kept a traverse profile showing in graphic form the status of the EVA in terms of time and life-support expendables. Using this profile, we were prepared to suggest modifications to the EVA plans if something unexpected happened or if the astronauts spent more time than planned at a given station. Time was always our enemy, and we knew the crews felt the same. How could they make the most of each minute yet not miss some important discovery? With the pictures coming back from the TV camera carried on the LRV, we were able to keep up with the crews' efforts and think ahead with them as to what should be the next priority. However, the crews always seemed to make the right decisions without many inputs from the "back room," a tribute to their training and dedication. There was seldom any second-guessing from those of us privileged to feel so close to the action even though we were 238,000 miles away from actually swinging a hammer or snapping a camera shutter. They were our surrogates in this inhospitable and strange land; we could only sit back and admire the job they were doing. Our job was to try to assimilate the information they were returning so as to arrive at the "big picture" of the Moon that their observations were starting to give us. We were careful not to interrupt the crews or burden them with unneeded questions.

The third EVA that the Field Geology Team had pleaded to retain became a quick trip to sample a northern crater (North Ray), about three miles away. Deposits around this crater were believed by many, but not by all, to be the best opportunity to collect volcanic samples. Time permitted only a few stops, but the traverse was made without incident and harvested many samples. The total weight of samples collected during the three EVAs was 211 pounds, a new record; the total distance traveled, sixteen miles, was slightly less than that recorded for *Apollo 15*.

Ken Mattingly, after the rocky start caused by the spurious gimbal motor readout, had successfully carried out his part of the science tasks. His sensors and cameras covered much of the Moon's surface between five degrees north and five degrees south of the lunar equator. Ken also spent as much time as possible making careful visual observations, which were valuable during the crew debriefings and in analyzing the data captured by the sensors.

Several months later, analysis of the *Apollo 16* samples showed that they, like most of the samples collected from the previous four missions, were breccias, the products of one or more cataclysmic events that showered the lunar surface with the debris from multiple impacts. None were volcanic in origin, however, and the Lunar Sample Preliminary Examination Team stated that "no evidence for lava flows or pyroclastic rocks was observed." This was an example of the pitfalls of trying to make photogeologic interpretations with no fieldwork to base them on. But we were learning with each mission how to improve our interpretations, and now we had an additional tool, the CSM data collected from lunar orbit that provided estimates of the composition of the lunar surface over wide areas.

With the *Apollo 12* ALSEP's third anniversary of uninterrupted operation approaching, I wrote a memo for Rocco Petrone's signature that was distributed to NASA management, including the new NASA administrator, James Fletcher, to report how well the ALSEP had performed.[4] We reminded those on the distribution that the original design goal was one year and that four experiments, the passive seismometer, Suprathermal Ion Detector Experiment, Solar Wind Spectrometer, and dust detector, were still operating normally and returning useful data. The Lunar Surface Magnetometer had operated successfully for two years, and only one experiment, the Cold Cathode Gauge, had failed immediately after activation.

The radioisotope thermoelectric generator was still putting out sixty-nine watts of power, four watts above predicted values for initial power output. The ALSEP central station had responded to over fifteen thousand commands during the three years and showed no sign of deterioration in spite of having experienced thirty-seven lunations that created temperature swings each time of over 500°F (-260°F to 270°F; by this time we had a more accurate measurement of surface temperatures). The *Apollo 12* ALSEP would continue to operate for almost five years longer.

It was 9:00 P.M. on December 6, 1972. *Apollo 17*, the last of the Apollo lunar flights, was on launch pad 39A at KSC with my good friend Jack Schmitt, Commander Gene Cernan, and CM pilot Ronald Evans strapped on board the command module, named *America*. As we had done for the last two flights, Rhett Turner and I were prepared to broadcast the launch for the Voice of America Worldwide Service from the press site about three miles west of the pad. To our right, along the raised, curved berm, were the large air-conditioned broadcast booths of CBS, ABC, and NBC, along with booths of other companies. If we used our binoculars we could see Walter Cronkite, Jules Bergman, and the other TV commentators looking out their picture windows at the brightly illuminated *Saturn V.* Voice of America was a bare-bones operation. We sat in the open swatting mosquitoes, last in line on the berm, on two folding chairs at a card table. Behind us, in what was once a two-wheeled camper trailer, was the engineer with all the electronic equipment. Later, after the lunar landing, along with my regular duties in the science support room at MSC, I was scheduled to make several broadcasts for the Spanish-language VOA using my rusty Colombian Spanish to explain what was happening—my last assignment for VOA.

Apollo 17 was scheduled for launch at 9:53 P.M., the first night launch for Apollo. So in addition to our excitement about this last launch and all it meant, we were looking forward to seeing and "feeling" the giant Saturn rocket roar off the pad. We could only guess at the visual impact, the spectacle of the world's largest firecracker lighting up the night sky. As we sat watching the brilliantly lit launch tower and rocket, the clouds that had partially obscured the sky were slowly dissipating; it had been predicted that if the sky was clear viewers as far as five hundred miles away might see the rocket as it streaked away to the east. Rhett had done his usual impeccable homework for the launch, the best of any reporter I knew, and he carried the audience along as the countdown proceeded, bringing me in as needed to provide some special insight. As we went through our rather informal script, Rhett would signal the engineer to play a previously recorded interview with Jack, Gene, or Ron or some other pertinent clip that would give interesting background about the mission. Illuminated in front of us, between our position and the pad, was the large digital countdown clock counting down the seconds. From time to time Chuck Hollingshead, the voice of Kennedy launch control, would interrupt our coverage with comments broadcast over the public address system.

Everything was proceeding normally until T minus thirty seconds, when without any warning a hold was announced. No immediate reason was given, and we were left, with all the rest of the commentators, to speculate on what was wrong. For the next twenty minutes we tried to make educated guesses, hoping that it was something minor and the countdown would soon resume. Finally Hollingshead came on the PA speaker and explained what caused the hold. The third-stage fuel tanks had not pressurized on schedule, and though a manual pressurization was attempted it was too late in the countdown and the automatic sequencer shut down the launch. It didn't sound serious, but it wasn't clear when the count would start again; we assumed it might take about an hour before the countdown would resume at T minus twenty-two minutes, the newly announced recycled starting point.

We underestimated. For the next two hours Rhett and I filled the airways with impromptu discussions of *Apollo 17* science and whatever other subjects we could think of. From time to time VOA would break away to provide news of the world and return to us, still sitting under the stars waiting for an announcement. Listening to those VOA tapes twenty-five years later is a real trip down memory lane. They record the late-breaking news and include the story that former president Harry Truman, age eighty-eight, was in critical condition in a Missouri hospital.

Finally the count resumed at T minus twenty-two minutes; it was held again at a planned point at T minus eight minutes to check that the pressurization trouble had been resolved before the countdown continued. At 12:33 A.M. on December 7, 1972, *Apollo 17* was launched; a historic date, one that will always be remembered along with another December 7 thirty-one years earlier. The liftoff was every bit as spectacular as we had hoped, lighting the night sky for miles around and pounding our bodies with the powerful low frequency reverberations that only a *Saturn V* launch produced. If you have witnessed a shuttle launch, multiply the effect by two. The crew of *Apollo 17* was on its way, and to top it off, we had survived two hours of unscheduled airtime! We packed our notes and left for Houston.

Apollo 17, after its prelaunch difficulties, was the most trouble free of any of the missions. All the pieces were falling neatly into place. We were definitely learning, but now we had no further chance to put this hard-won education to use. The landing site, Taurus-Littrow, almost as far north as the *Apollo 15* landing, was on the edge of Mare Serenitatis, to an Earth observer the right edge of the man in the Moon's right eye. The landing would be the most difficult maneuver of any flight yet, requiring Cernan to come in over the Taurus Mountains, 6,500 feet high, descend steeply into a narrow valley, and land between the bases of two mountains.

Taurus-Littrow was selected for the final Apollo landing for several reasons. During the *Apollo 15* mission Al Worden had observed that this area was covered by a mantle that looked darker than other parts of the lunar surface. His observations seemed to be confirmed by Lunar Orbiter photography and photographs taken from orbit during other Apollo missions. The promise of finding "recent" volcanism raised its head again. Would this site provide samples that would confirm an epoch of late lunar volcanic activity? Samples from the Taurus Mountains were also of great interest. Would they be similar to or different from the highlands samples collected on *Apollo 16*?

As a far northern and eastern landing site, it had value for several of the ALSEP experiments, in particular for the Lunar Atmospheric Composition (LACE) experiment (the passive seismometer was not carried on this mission), which would provide better data if separated in distance from the other ALSEPs, which were still sending measurements. LACE, a miniature mass spectrometer, was a more sophisticated version of the Cold Cathode Gauge experiment deployed on missions 12, 14, and 15 to detect the tenuous lunar atmosphere. We also had two new surface experiments on the mission, Surface Electrical Properties (SEP) and the traverse gravimeter, which, along with the portable magnetometer that flew on *Apollo 16*, would be operated by the astronauts during the LRV traverses. These last three experiments were expected to provide important information on the subsurface structure of the valley at the base of the Taurus Mountains.

While missions were under way, my job at MSC included manning a console in the Science Support Room and taking part in the discussions that would fill the exciting hours while the astronauts were on the lunar surface. Occasionally I would spell the headquarters duty officer in the Mission Operations Control Room, the latter largely a ceremonial duty if the mission was proceeding according to plan. In addition, I would participate in briefings with VOA and other news organizations. Although the word "spin" had yet to be applied to government briefings, that was part of our approach. If something in the mission timeline didn't go according to the material passed out to the media before the mission, we would be interrogated at each of the daily updates held in the MSC auditorium. No question was off limits, and some would be off the wall, reflecting the media's understanding, or misunderstanding, of what was going on.

We all had our preferred media person to talk to off line, someone we knew from experience would tell the story reasonably straight and get the facts right. My favorite was Donald Kirkman of Scripps-Howard. I tried to avoid Thomas O'Toole of the *Washington Post*, with some success, for he seemed to be always looking for the negative side of events and could usually be counted on, at some point during a mission, to misinterpret an important story. We would feed trusted reporters tidbits of insider information so that their stories would be more informative or have a little more punch than their competitors'.

Just before the crew achieved lunar orbit, their discarded SIVB stage hit the Moon about 525 miles west of the *Apollo 16* ALSEP, and the impact was recorded by all four passive seismometers that were still operating from the earlier missions. This time there were no problems in lunar orbit after separation of the LM and CSM, and Cernan accomplished the landing after taking over control from the autopilot and set the lunar module, *Challenger,* down in the rock-strewn valley between the North and South Massifs.

Once on the surface, Cernan and Schmitt, the last men to set foot on the Moon for what has turned out to be three decades and counting, went energetically about their business. The crew had trained hard, and we could sense from Cernan's descriptions that he was not about to be outshone by his geologist teammate when it came to conducting their surface studies. They described the sight that confronted them as "spectacular." I have probably overused that word in this story as much as the astronauts did during the missions, but it is the best adjective I know to describe the views we could see from the TV images and later from the many excellent photos they returned. To the north, less than seven miles away, the mountains rose almost perpendicular from their base toward the black sky. To the southwest, again less than seven miles away, lay the South Massif, equally imposing if not quite as steep as the mountains to the north. TV pictures captured the landscape clearly, and in the SSR we could only wonder at our audacity in asking the crew to land in such close quarters.

Apollo 17 reconfirmed the targeting ability of the MSC engineers. They brought Cernan and Schmitt to the precise point where Cernan was scheduled to take control, and he then successfully demonstrated his landing skills. With this experience it seems certain that if missions had been scheduled after *Apollo 17* we could have persuaded management to agree to landings at important sites such as the central peaks or rims of Copernicus and Tycho. Future lunar explorers, undoubtedly piloting spacecraft with greater capabilities, will find safe landing sites almost anywhere on the Moon, including the farside!

Apollo 17's first EVA began with the removal of the LRV from its stowage bay on the descent stage and the erection of the TV high gain antenna. Thereafter we had good TV coverage of the landing site and the astronauts deploying the ALSEP. They drilled three holes, two for the heat flow experiment and the third to recover a ten-foot core. ALSEP began transmitting data as soon as it was activated, and the star of the experiments, the Surface Gravimeter, provided strong signals. Little did we know at this point that there was a major problem, as described in chapter 7. After finishing these tasks, the astronauts still had time to take a short (about half a mile) ride to the south to collect samples near a small crater named Steno. During this traverse they also took Traverse Gravimeter and SEP readings and left two explosive charges to be detonated later to provide signals for the Seismic Profiling experiment.

A few words about the explosive packages that were an integral part of the Seismic Profiling experiment. Commencing with the design of the active seismic experiment, carried first on *Apollo 14*, we went through an extensive review and certification of the explosives used with that experiment and the Seismic Profiling experiment. Some at NASA were not happy about carrying live explosives on the LM, so our test procedures were carefully monitored. We had to prove beyond any doubt that there could be no accidental firing of the charges. Petrone, especially, followed the certification process from beginning to end and witnessed some of the field tests.

Fortunately this experiment was not the only place explosives were used during the mission, starting with the separation of the launch escape tower from the CSM and progressing through the individual rocket stages, where explosive squibs were used to separate some of the stages during flight. We benefited from all the work that went into qualifying these explosives and designed our charges using aspects of these proven designs. The biggest fear, of course, was that an inadvertent firing command, short circuit, or other accident might trigger the explosives, either while they were stowed on the LM or while the astronauts were setting up the experiments.

Because the astronauts would hand carry the explosives on the lunar surface, every firing circuit had either double or triple safety redundancy before the firing commands could activate the charges. For the Seismic Profiling experiment, the arming sequence was as follows: Each explosive package had three pull rings on top. Pulling ring one started the safe/arm timer. Pulling ring two, and rotating it ninety degrees, released the safe/arm slide to start the mechanical timer. Pulling ring three cleared the firing pin and placed a thermal battery timer on standby until a coded signal was received from the ALSEP central station, the preferred way to set off the charges. In case ALSEP commands weren't received, the mechanical timers were preset for periods from 89.75 to 92.75 hours after activation, well after the astronauts left the lunar surface. Each charge package had an antenna that would receive the initiation signal from the central station to start the firing sequence, at which point there was a two-minute window in which to receive the coded firing signal.

Sound complicated? It was. There was a running joke that with all the safety features we would be lucky to get even one to fire. But all eight charges were fired successfully after the astronauts departed, and the experiment's four geophones recorded the explosions, providing information about the upper mile and a half of the Moon's subsurface. The LM ascent stage impact, about seven miles southwest of the landing site, with its higher energy input, allowed Bob Kovach, the PI, to improve his measurements and estimates of seismic velocities to a depth of three miles below the surface.

The second EVA, the longest of the three, went almost due west and then swung southwest to study and sample the South Massif. Cernan and Schmitt made several stops along the way, including sampling the rim of a fairly large crater called Camelot. On finishing their work at the crater, they drove up the low scarp that separated the valley from the massif and sampled the boulders at the base of the massif. They then returned to the LM by a different route that took them farther north on the valley floor and included seven more sampling stops. One of these stops, at Shorty Crater, was to sample the dark halo material surrounding the crater that could be seen on Lunar Orbiter photographs. Worden had reported that he could see a color difference from orbit. Once again we hoped the dark material would be "recent" volcanic deposits as was predicted (incorrectly) at the *Apollo 16* landing site. What they found, to the great excitement of both astronauts, was an orange and red soil interspersed with darker and lighter soils. Schmitt thought they had found the elusive recent volcanic vent. Returning to the LM at the end of the EVA with their find, they placed three Seismic Profiling explosive packages at varying distances from the ALSEP central station and took a series of Traverse Gravimeter and SEP readings.

The third EVA traverse was made toward the east and then turned north to sample the North Massif and the intervening darker plains material, also interpreted as volcanic mantling material. Traverse Gravimeter readings were made on this EVA, and the final three Seismic Profiling explosive charges were placed on the surface at intervals along the route. Because the SEP receiver overheated, no data were collected by this experiment during the third EVA.

By now, after analyzing the signals coming back for almost two days, we realized that the Surface Gravimeter was not responding correctly. We cut the third EVA short to allow Schmitt to go back and rebalance the gravimeter's movable beam. This last attempt to improve the experiment's response while the astronauts were still on the Moon also failed. Something was wrong, but we weren't sure what it was and couldn't find a solution. Closing out the final Apollo lunar surface EVA on a somewhat dismal note because of the gravimeter problem, Cernan removed the neutron probe from the core hole for analysis back on Earth and climbed back into the LM.

With the crew of *Apollo 17* safely on the Moon, Dale Myers and Rocco Petrone released the final "Apollo Program Plan."[5] It covered all the remaining *Apollo 17* activities and those actions necessary to close out the Apollo program. Among other notes of finality, it stated: "All basic hardware procurement for the Apollo Program has been accomplished." The nation would not purchase any more awe-inspiring *Saturn V*s or superbly engineered CSMs and LMs. The schedule in the plan showed a transfer of responsibility for common Apollo-Skylab activities to the Skylab Program Office, the next approved program, by the middle of FY 1973. Beginning on the same date, the remaining Apollo lunar science activities, mostly monitoring the ALSEPs and publishing results, would be undertaken by the Office of Space Science and Applications. Although *Apollo 17*, and for that matter the entire Apollo program, had achieved all its objectives and more, this final Apollo plan ended an era with the sour taste of a great opportunity lost through lack of national leadership. It was an era that had begun with great expectations of conquering new worlds.

After almost seventy-five hours on the lunar surface, twenty-three of them spent outside the LM on the three EVAs (another new record), Cernan and Schmitt lifted off to rendezvous with Ron Evans, carrying with them the 243 pounds of samples they had collected at Taurus-Littrow during traverses that covered more than twenty-one miles. The LM was jettisoned, and this time the impact occurred just west of the landing site, some 475 miles east of the *Apollo 15* ALSEP. Its impact was recorded by all four of the previously deployed passive seismometers and the Seismic Profiling experiment deployed on this mission.

While Cernan and Schmitt were on the surface, Ron Evans had been conducting the experiments assigned to him. He had also undertaken some experiments beginning with the translunar coast phase and would continue making measurements during the return home, almost until reentry. The high activity period, however, was while he was in lunar orbit, where he conducted a suite of experiments similar to those of the previous two missions. Once Cernan and Schmitt were on board and they were on their way back to Earth, Evans carried out an in-flight EVA when he retrieved the film canisters and other data from the SIM bay. Splashdown and recovery were uneventful.

Project Apollo was now a new chapter in the history books. Even with a few glitches, the flight of *Apollo 17* had been the most successful mission from a scientific viewpoint. An enormous treasure trove of lunar samples was in the vaults at the Lunar Receiving Laboratory awaiting study. The seismic and laser corner reflector networks were already returning exciting information, as were many other ALSEP experiments. ALSEP central stations were performing up to or beyond their original design goals. But political and public apathy had set in long before the launch of *Apollo 17,* and the scientific results alone couldn't convince the decision makers to add more missions. Those of us still working in the Apollo Program Office faced the dismal task of mopping up and closing down an unequaled undertaking. Many of my coworkers had already begun to drift away to other NASA offices or to new work in or outside the government.

Our dreams for lunar exploration never went away; we always hoped that Congress and the Nixon administration would see the error of their ways and provide the funding to reinstate our post-Apollo plans. But in spite of the surprising discoveries made by *Apollo 11* and the missions that followed, no national commitment was forthcoming, and the Apollo hardware remaining after *Apollo 17* was never used for its original purpose. Some was used for Skylab, some for Apollo-Soyuz, and the other items are lying ignominiously on the ground at museums, like tethered Gullivers, as reminders to millions of visitors each year of Apollo's magnitude—and perhaps, to some, of opportunities lost. When will man again set foot on the Moon? Or will we bypass the Moon and go directly to Mars? Or will we stay earthbound or in near-Earth orbit for generations?

One afternoon, walking between my office at L'Enfant Plaza and the NASA offices at 600 Independence Avenue, I met some of my former Advanced Manned Missions and Apollo colleagues who had recently been assigned to potential future manned space flight programs. We talked briefly about the uncertainties surrounding these programs (none had been officially blessed as the successor to Apollo except for the short-term Skylab and Apollo-Soyuz programs) and discussed where I might find a new job. I was struck by their lack of enthusiasm and their pessimism about their new work as we discussed NASA's future. It seemed as if almost overnight this marvelous can-do agency had grown old and lost its way. Gray heads were beginning to predominate in all the offices. Instead of looking toward an exciting future, everyone seemed to be scrambling to hang on and find a place to roost. I decided at that moment that it was time to leave NASA and find a new program I could devote my energies to.

Following the lead of Ed Davin and Dick Green, I sent an application to the National Science Foundation and was hired immediately by a new organization called Research Applied to National Needs, which was undertaking research on a wide spectrum of new technologies. Thus began a new career, but never again would I experience the excitement and the sense of achievement that came with being a small part of Project Apollo.

The Legacy of Apollo

One of the questions most frequently asked at the end of Project Apollo, and even today, is What did we learn? It's a good question. It would often be followed by other, unanswerable questions. Was the project worth the cost? Wouldn't we have been better off spending the billions of dollars on X, Y, or Z?

Addressing the unanswerable questions first from the perspective of science, it is difficult to calculate the part of Apollo's cost that funded scientific experiments for the Apollo flights, because the many components that made up Apollo science were carried in different parts of the NASA budget. Should it include the salaries and overhead for all the civil servants involved? Should it include the support contractors' costs for those who worked at NASA centers and headquarters and were involved in the planning and development of the science? How about facilities such as the Lunar Receiving Laboratory? The LRL cost over $16 million to build and equip, plus additional operating expenses during the missions. Add to that sum the $19 million given to the sample analysis principal investigators, and the expenses for quarantine and sample analysis alone total over $35 million.

Considering how the post-Apollo studies contributed to Apollo science, should any of those costs be added to the total? Should all the advisory committees and summer conferences that were funded by NASA? They contributed important advice and helped us select the experiments included on the missions. And of course there were the costs associated with integrating experiments on the lunar module and the command and service module and training the astronauts in their use and deployment. Finally, there is the cost associated with experiment data reduction. Calculating an accurate sum for all these activities is probably impossible at this late date, and the items mentioned probably overlook other costs that would contribute to a grand total.

NASA bookkeeping, like that of many government agencies and cabinet departments, used a document called a program operating plan. The POPs categorized expenditures by program, and within each program the expenditures were further delineated by a work breakdown structure or, in simpler language, an item-by-item accounting. These terms are important only to show that there was great rigor in keeping track of taxpayers' dollars. Each office and center within NASA kept these records, and they were compiled and reviewed by the NASA headquarters Office of Programing. This office not only kept track of expenditures but was also the focal point for preparing each year's budget requests to the Bureau of the Budget and its successor, the Office of Management and Budget, and then with other senior management presenting and defending the budget before Congress. The name of this office changed through time, but the men who ran it, such as DeMarquis "Dee" Wyatt and William Lilly, were both feared and admired because of their power to approve or disapprove program requests.

In the Apollo era, budget control was a hallmark of NASA, and discussion of the flow of funds for all programs took up a major part of Jim Webb's and Bob Seamans's monthly program reviews with the associate administrators and lesser managers. Program managers were expected to keep their books up to date and in good order. Any deviations from approved schedules and budgets during an individual program's lifetime had to be fully explained and justified, on pain of strong reprimand or even demotion or removal. Considering the uniqueness of this new frontier and the challenges it represented, only a few large overruns occurred. The Surveyor program was an example. The problems that have plagued the International Space Station in recent years, including schedule delays and large cost overruns, would not have been tolerated in the early days of NASA by either NASA management or Congress. But that is a story for another day.

My "hard" number for Apollo science includes estimates of the components listed above and is based on reviews of microfiche records, internal memos, and POPs, of which the last one I had access to was POP 72-1C.[1] There may have been later POPs covering Apollo science, but this one showed closeout costs for the last four years of the surface and orbital science programs and contractor manpower ramping down toward zero. The total reported in POP 72-1C was

$150,000,000—a nice round number, but I believe it was understated. In a memo to NASA Public Affairs, responding to a request for the costs of the *Apollo 15* experiments, we made an estimate of $36 million.[2] In another memo, this time to the NASA budget office in March 1972, we estimated a total expenditure of $85 million for the *Apollo 16* and *Apollo 17* missions, a total of $121 million for just the J missions.[3] This indicates to me that the 1972 POP did not include important pieces; however, you can't tell the basis for the number— what was included or what may have been left out. In William David Compton's history of Apollo, *Where No Man Has Gone Before,* he indicates that slightly more than $218 million was spent on science payloads.[4] But once again it is not completely clear what this number represents.

An estimate I made at the end of the program was $265 million, based on calculating the cost of each experiment and adding other related contractor costs available at the time. But that estimate did not include some of the items described earlier. I now believe the total would come close to $350 million in 1972 dollars, not including civil service salaries and benefits. If you accept this number, the science piece of Apollo was about 1.5 percent of the total $25 billion spent.

If we factor into the $25 billion the national prestige value of being the premier spacefaring nation, the excitement of visiting a new world, the knowledge gained about the Moon and Earth, and the advanced technologies that resulted from Apollo (some call it spin-off), we can try to answer the unanswerable. Was it a bargain, money well spent, or money wasted? My judgment: unequivocally a bargain!

What did we learn? remains an important question because as students of the Moon continue to examine material brought back during the Apollo missions, fresh results are still coming in. New information from the recently completed Clementine and Lunar Prospector missions adds to our knowledge and clarifies or extends the Apollo results. More than 1,100 abstracts were received for the Thirtieth Lunar and Planetary Science Conference held in Houston in 1999, approximately one-quarter dealing with lunar subjects, attesting to the continued interest in Moon-related studies. Programs to return to the Moon, based on a desire to learn more about our nearest celestial neighbor and perhaps begin to exploit its resources, are constantly proposed.

If, as many of us who worked on Apollo fervently hope, the United States (perhaps in concert with other nations) mounts another Apollo-type project to send astronauts back to the Moon or on to Mars, then we must be prepared to justify and explain to the American public the benefits of spending a nontrivial amount of the national budget on such undertakings. At the moment NASA management does not support going back to the Moon, on manned or unmanned missions. In spite of the interest in recent Mars missions, sustaining public support for extended Mars exploration will be difficult. To the casual observer, or the average taxpayer, one picture of a Mars landscape will look much like the last one, even if it includes an astronaut holding a rock, pointing at a mountain, or riding around on some strange-looking vehicle.

If a political objective is not the driving force at the time the debate begins, as it was at the start of the Project Apollo, then we must be able to predict scientific and economic benefits of value to those on Earth. Such predictions will be difficult to make unless we can provide a connection to what we gained from the Apollo Moon landings and extrapolate this knowledge in a rational way to these new projects. Intellectual adventures will not suffice, even if one believes that the thrill of exploring new lands still survives in our species.

Briefly, here is a summary of the findings and lessons learned from Apollo and the Lunar Orbiter missions that immediately preceded Apollo. A few of Lunar Orbiter's contributions are discussed briefly, and the notes list references that provide details on this program. From all the missions and other programs, such as the radar studies conducted from Arecibo in Puerto Rico, by the end of Apollo we came to a new understanding of the Moon.

Lunar Orbiter's comprehensive, high resolution coverage of the Moon's surface allowed lunar students to expand their understanding of the Moon in significant ways.[5] For example, the higher resolution pictures permitted the United States Geological Survey lunar mappers to refine the geological studies they had been making for the previous four or five years based on telescopic observations. Before Lunar Orbiter returned its magnificent photographs, geological formations mapped by USGS workers were distinguished by such characteristics as subtle differences in albedo (reflective power), surface roughness, and crater counts. With higher resolution Lunar Orbiter photographs in hand, the quality and speed of their work increased. The validity of their interpretations would have to await the additional information to be returned by Apollo missions.

In retrospect, one would have to give USGS a good grade (perhaps a B+) for its early efforts. Physical differences were no doubt present; what they represented was difficult to predict. Forced to make interpretations based on these subtle distinctions, and working under the great disadvantage of not having material in hand that represented their mapped formations, some overestimated the complexity of the Moon's surface. Perhaps the best illustration was the view that many places on the Moon exhibited volcanic features such as cinder cones. The *Apollo 16* landing site, selected in part to permit the astronauts to sample this type of feature, returned mostly breccias and no volcanic ejecta. However, the famous "orange soil" found at the *Apollo 17* site is interpreted to mean that it was formed during lava fountaining from a volcanic vent, but almost 4 billion years ago. No traces of "recent" volcanism were found. Nothing significant seems to have occurred on the Moon for at least the past 50 to 100 million years except for random impacts.

Lunar Orbiter's farside coverage allowed the USGS mappers to extrapolate their extensive nearside studies to this perpetually hidden face. Its appearance, highly cratered and without the vast, smooth maria common on the nearside, differed from the face of the Moon that had been studied for centuries. It looked much more like the nearside lunar highlands. With a few exceptions, such as the large crater named Tsiolkovsky and the Mare Moscoviense basin, the large farside impacts had not filled with marelike flows as had many of the large nearside impacts. This difference was attributed to the pull of Earth's gravity, with the nearside being much more strongly influenced during the early history of the Moon than the farside, thus allowing lunar basalts to fill the low-lying nearside basins. This conclusion supported the belief that early in its formation the Moon had become locked into its present orbit, with its orbital rotation around its axis of twenty-eight days equaling its orbital period around the Earth.

Lunar Orbiter also permitted a more detailed analysis of the Moon's gravitational field and the irregularities in the field. Its ability to provide this information had been proposed by Gordon McDonald. By closely tracking each spacecraft's orbital path and calculating how it differed from the path that would be expected if the Moon's gravity field were uniform, lunar geodesists were able to accurately plot, for the first time, the figure of the Moon. This close tracking led to the discovery of the "mascons" (mass concentrations) mentioned earlier. Deviations from Lunar Orbiter's calculated flight path suggested that material denser than the surrounding terrain formed the widely scattered mascons. These data were upgraded by tracking the CSM on each Apollo mission, and more recently they have been refined by tracking the orbits of Clementine and Lunar Prospector.

Finding the mascons has important geological and geophysical implications that should contribute to deciphering the Moon's early history. In addition, knowing the mascons' positions will be especially useful when we return to the Moon with either manned or unmanned missions, because it will allow us to program the landers to arrive precisely at their designated landing sites. But Lunar Orbiter, Apollo, and recent missions could not tell us what the mascons are or how they were formed. Resolving these questions will have to await additional measurements made by the next generation of spacecraft.

Ranger, Surveyor, Lunar Orbiter, and Apollo put to rest for most lunar students the question of the origin of almost all lunar craters: they were formed by impacts. This knowledge has led us to look at the Earth's history in a new light.[6] Before the Apollo landings, most Earth scientists believed that in its earliest history the Earth had witnessed a period of intense infall of large and small planetesimals, meteorites, and other debris from a newly forming solar system. Little direct evidence of this epoch could be found in the geological record, and until recently only a few impacts had been positively identified and studied. The rate at which these large and small impacts bombarded the Earth was pure speculation, but those who studied these features believed large impacts were probably common.

Today some terrestrial features and events that had previously been difficult to explain are being attributed to large impacts. The most fully reported event of this type provides an explanation for the disappearance of the dinosaurs and many other species of animals and plants at the end of the Cretaceous period, some 65 million years ago. Proposed in 1980 by Luis Alvarez, his son, and several other researchers,[7] the theory was based on the discovery in Italy of a thin rock formation, enriched with the element iridium, at the Cretaceous/ Tertiary geological boundary.

They concluded that the best explanation for this anomaly was that a large object composed of material containing a high percentage of iridium, a common constituent of certain types of meteorites but not common in Earth rocks, had struck the Earth at this precise time. Debris from this impact spread over a large portion of the Earth's surface and was deposited as a thin layer that included the formation discovered in Italy. This proposal was met with great skepticism by many in the scientific community, but some, including Gene

Shoemaker and others who had been involved in Apollo science, supported the idea, knowing that large impacts had affected the Moon's history. This Cretaceous impact has been confirmed, and through the work of many scientists, the probable impact site has now been located on the edge of the Yucatán peninsula in the Gulf of Mexico. Whether it led to the species extinctions observed at the end of the Cretaceous period is still being debated.

The Apollo program's emphasis on understanding impact craters spurred the search for and discovery of other large Earth impacts. For example, an ancient impact crater has been found in Texas, at Sierra Madera; another underlies Chesapeake Bay; and a buried crater in southeastern Nevada is believed to have created the Alamo breccias. The identification of impact events in the geologic past has accelerated as our diagnostic techniques have improved. Australia has been especially productive for the study of impact craters because much of its surface has remained relatively undisturbed for millions of years. It was while undertaking such a study that Gene Shoemaker met his untimely death.

These discoveries have led to a related field of study, tracking objects orbiting near the Earth and crossing the Earth's orbit (hundreds are now known) that might strike the Earth in the future. Today, if such an object took aim at Earth, it could not be avoided. If another object the size of the one that hit the Earth at the end of the Cretaceous period were to strike the planet, it would trigger a series of events with unimaginable consequences. But not much is being done to prepare for such an admittedly low-probability event. Some believe we could avoid such an impact, if it was predicted, by developing an early warning system that would track large meteors or asteroids and then deflect them with missiles. The study of impacts on the Earth and Moon has resulted in a model that predicts the frequency of impacts on the Earth. This model suggests that a large impact occurs approximately once every 50,000 to 100,000 years. Perhaps this knowledge will motivate world governments to work together for a solution that will prevent such a catastrophe.

Although a relatively small event when it occurred, the Meteor Crater impact has been dated at approximately 50,000 years ago. It undoubtedly was a devastating blow for a large region surrounding the impact point, creating ground tremors and clouds of dust and debris that would have extended over hundreds of square miles. At that time the only casualties may have been a few mastodons and other wildlife. If such an event occurred today, Flagstaff and other nearby towns would probably be destroyed, and cities as far away as Tucson and Phoenix would feel its effects. Are we due for another big impact—soon? The model suggests we may be.

By the end of the Apollo missions, the six successful landings and their predecessors had returned a wealth of new information about the Moon. Before the landing missions, *Apollo 8* and *Apollo 10* traveled to the Moon but did not land. *Apollo 8* was the historic mission that orbited the Moon at Christmas 1968, with men being captured for the first time by the gravity field of a "planet" other than the Earth. Although we had in hand excellent close-up photographs from Lunar Orbiter, this was the first time men were able to view the Moon at close range.

The lunar farside especially impressed the crew of *Apollo 8;* Frank Borman, Jim Lovell, and William Anders reported seeing a jumble of craters on top of craters. Orbiting sixty-nine miles above the surface, they described the Moon during their Christmas Eve greeting to those back on Earth as "a vast, lonely, forbidding . . . expanse, . . . it certainly would not appear to be a very inviting place to live or work."[8] Fortunately we were not going to try to make an Apollo landing on that "forbidding" farside terrain. *Apollo 8* also gave us the first views of our home planet from a great distance away; the Earth was described as an oasis, isolated in the emptiness of space. Some have credited this dramatic view of Earth with imparting a new awareness of how unique our planet is and how important it is to protect its fragile environment—an unexpected bonus from the Apollo program.

Apollo 9, launched in March 1969, was the first test of all the Apollo hardware working together as it would for a Moon mission except that the crew and the equipment never left Earth orbit. It was followed two months later by *Apollo 10,* a dress rehearsal for the first attempt to land on the Moon. The crew of Tom Stafford, Gene Cernan, and John Young would perform all the complicated maneuvers required of a landing mission except for the most crucial—the actual landing. Stafford and Cernan would separate from the CSM in lunar orbit, descend to less than ten miles above the lunar surface, jettison the landing stage, activate the LM ascent engine, and rendezvous with the CSM. Close, but oh so far from making history. In addition to testing all the elements leading up to a landing, they proved the accuracy of Apollo targeting and the astronauts' ability to see their landing point and observe potential hazards at a site similar to that expected for the first landing in the Sea of Tranquility. *Apollo 8* and

Apollo 10's reconnaissance also confirmed what had been seen in the Lunar Orbiter photographs: smooth landing areas were available in the "Apollo zone."

Cernan and Young would get another chance to perform a Moon landing; Stafford is the only man to get within ten miles of the Moon and never land. I'm sure he would gladly forgo that honor for the thrill of having kicked a little Moon dust. Two months later the first landing would take place. Many other Apollo prime and backup crew members—Walter Schirra, Donn Eisele, Walt Cunningham, Jim McDivitt, Gordon Cooper, Joe Engel, and Rusty Schweickart—would suffer the disappointment of being selected to test the Apollo hardware but never getting to the Moon. But without their key roles and dedication the Moon landings could not have been undertaken.

By the time *Apollo 11* splashed down we had developed the routine by which the science results would be processed and disseminated. The astronauts would be picked up by a navy helicopter operating from an aircraft carrier, transferred to a specially designed trailer on the carrier, and flown back to Houston to be placed in quarantine in the LRL. The samples, film, and other data would be removed from the command module and flown to the LRL in their own aircraft. Once in the LRL, the astronauts would be debriefed by a team of scientists and engineers while the samples were unpacked, examined, and cataloged and the photographs were developed. In the meantime, we would be receiving data from the instruments left on the Moon.

This routine was followed for all the missions, with the major difference that after *Apollo 14* the astronauts and the samples no longer had to spend time in quarantine and the debriefings became much more relaxed and easier to carry out. Without an intervening barrier, we could question the crews much more directly as we tried to piece together all they had done. This was an important change, because the last three missions were more complex and the astronauts' recollections more valuable in reconstructing their long traverses on the lunar roving vehicles. I will describe only how we debriefed the *Apollo 11* crew and studied the first samples, but I will include results from all the missions to explain what Apollo taught us.

The *Apollo 11* astronauts had their first science debriefing on August 6, 1969. Before that debriefing the Manned Spacecraft Center engineers had reviewed the nuts and bolts of the mission—the "technical debriefings"—going over those aspects of their flight that might affect the success of the next mission. Although all spacecraft systems were carefully monitored by telemetry, with records kept of all discrepancies, the astronauts' answers to questions would often clear up troubling inconsistencies or uncertainties in the records. Neil Armstrong's and Buzz Aldrin's descriptions of their landing maneuvers and their difficulties in finding a good landing site were examples of how their experience contributed to improving the landing sequence for *Apollo 12* and later missions. All members of the crew of *Apollo 12*—Pete Conrad, Dick Gordon, and Al Bean—as well as the backup crew led by Dave Scott, were the most interested participants in these debriefings. Pete must have gained valuable knowledge, because he landed right on the money, within easy walking distance of his target, *Surveyor 3*.

Although I was invited to the science debriefing, I sat in one of the back rows while a few designated individuals, including Don Wise and O. B. O'Bryant from our headquarters office, were allowed to ask questions. It was a strange scene for such a momentous occasion, with the questioners and hangers-on peering at the three astronauts, who sat behind a brightly lit picture window like animals in a zoo. Unfortunately the transcript of the debriefing does not always identify the questioner, but Gordon Swann and Henry Holt of USGS and MSC's Gene Simmons, among others, covered all the important questions relating to the astronauts' surface observations, especially those that might affect what was planned for the next mission.[9]

Everyone involved in this debriefing, and in debriefings for later missions, came away with a great admiration for the astronauts' powers of observation and recall. When these traits were added to their innate resourcefulness and doggedness in following and going beyond their ambitious timelines, every possible ounce of science was gleaned from the missions in spite of the constraints they were working under. Some might take issue with that statement, but I believe it is true; the training and simulation had paid off beyond our expectations. Explorers of all generations have been eulogized for daring to take chances beyond the imagination of the ordinary person—for the astronauts it was called "the right stuff." However you wish to identify this urge to explore, it was undeniably present in these first voyagers beyond the friendly Earth who risked never returning to their home planet, a danger never before faced by explorers.

The *Apollo 11* science debriefing was our first chance to talk directly to the crew after their return from the Moon. After two weeks of isolation, interrogated every day by engineers and technicians, the astronauts were in a surprisingly good mood. From time to time one could sense a little irritation at questions that were repetitive or trivial, but all in all there was great cooperation, and we gained much from listening to their firsthand observations while they were still fresh in their minds. Their photographs had been developed and were available to supplement the discussion, as was the preliminary traverse map of the landing site. When necessary, the astronauts used large pads and marker pens to illustrate their answers, and Armstrong, especially, took advantage of these aids. (Were these unique drawings preserved for future generations of historians?)

One of many exchanges was particularly interesting. While in lunar orbit, before beginning their return to Earth, the astronauts were asked to look toward the crater Aristarchus and describe it. Although Aristarchus was just on the horizon and at the limit of their view, Armstrong reported that he thought he saw fluorescence in that region. This announcement caused some stir: Was he observing some lunar transient phenomenon like that described in chapter 2? Now, during the debriefing, he went into more detail and modified his observation. Although he described the general area as the brightest spot he could see, he could not confirm that it was Aristarchus itself that was causing the bright reflection, and he did not "mean to imply that it was self-illuminated." The unusually bright appearance of the Aristarchus region to the crew in orbit reinforced the belief that it might be the site of recent activity on the lunar surface. With their many other observations and much hard work, the crew of Apollo 11 had opened the door to a new era in planetary science.

The possibility of bringing some deadly unknown disease to Earth in the samples or by an infected crewman led to building the Lunar Receiving Laboratory. Strict protocols had been developed to guard against these risks. Some in the media latched on to this potential hazard, attempting to fan the public's fears of some catastrophic invasion. One month before the Apollo 11 liftoff, a media briefing was held in Washington to describe the details of the mission and, we hoped, allay any fears that the first landing and return from the Moon's surface posed any danger to life on Earth.[10] The final portion of the briefing was conducted by Air Force colonel John Pickering, who had served on the Interagency Committee on Back Contamination and now held the title director of lunar receiving operations at the Office of Manned Space Flight. He went to great lengths to describe the procedures that would be followed, from collecting and packaging the samples on the Moon through recovery and transport of the samples and astronauts to the LRL and eventual release of both at the end of the quarantine period. He even went so far as to include in the press handout a copy of the LRL biological certification signed by Dr. David J. Sencer of the United States Public Health Service, chairman of the Interagency Committee, to prove that all precautions had been taken. This openness and attention to detail defused this issue for most of the media, and it never surfaced again as a major public concern. However, managers of Mars missions that will return samples to Earth should head off the potential negative exploitation of this issue by being open and detailing the steps that will be taken to guard against alien organisms.

Let me illustrate how seriously the quarantine protocols were followed. During the preliminary study of the *Apollo 11* samples, a technician was cataloging a sample in an isolation chamber glove box that operated under negative atmospheric pressure to avoid any leakage into the LRL when one of his gloves ruptured, exposing him to the sample. This man and another working near him were immediately placed in quarantine in the LRL with the three astronauts. One lemon-sized rock was carefully sterilized and taken out of the isolation chamber and given to the Lunar Sample Preliminary Examination Team (LSPET) so they could hold it. Cliff Frondel, a member of the team, was quoted as saying, "It was a great thing to look at this stuff that people had been speculating about for millennia, and here it was in our hands. . . . It was a hell of a thrill."[11]

To determine if there were possible life-threatening forms ("replicating species") in the samples, ten species of animals were exposed to lunar material for twenty-eight days, either through inoculation or in their food. Four control groups were exposed in a similar fashion to nonlunar material for the same period. These animals included paramecia, planarians, shrimps, oysters, cockroaches, and houseflies. One might wonder if the testers, fourteen scientists called the Lower Animal Test Team, had any second thoughts about including cockroaches, insects that seem to be indestructible and have survived 200 million years of evolution essentially unchanged. Why would a little Moon dust hurt them, regardless of what it contained? The cockroaches and the astronauts cooped up together in the LRL became the basis of many jokes.

During the quarantine period, these ten species, living in small aquariums or jars and bowls inside the LRL, were carefully monitored for any suspicious behavior or a sudden desire to go to the special heaven reserved for them.

Nothing much unusual was observed; only the oysters, both those exposed to lunar material and the control groups, seemed to have a higher than expected death rate. This was attributed to conducting the tests during their normal spawning season, which apparently is stressful to romantic oysters. But as in all true scientific inquiry, one strange behavior was noted: planarians exposed to heat-sterilized lunar material swam at the surface of their bowl more frequently than the control groups.[12] The reason was unknown.

In addition to the study of "lower animals," similar tests were conducted on mice and quail. After four weeks of exposure to lunar material, 230 mice and 120 quail were autopsied by another team. Like the "lower animals," the inoculated mice were found to be normal, and the quail that had lunar soil mixed in their feed showed no adverse reactions. The reports from these teams were greeted with a sigh of relief from all 142 sample PIs and the scores of coinvestigators waiting anxiously to receive their allocated portion of the returned samples and get on with their analyses. If some pathogen had been found, we might still be waiting to study *Apollo 11*'s lunar treasure. The samples were declared safe for distribution around the world and were released on September 12, 1969.

The time between the conclusion of the contamination tests on August 22, 1969, and the release of the samples twenty-one days later was spent in preliminary analyses and preparing the specific types of samples required by each of the sample PIs. On August 27 the Lunar Sample Analysis Planning Team (LSAPT), chaired by Gene Simmons, issued a final internal "summary report" on its findings from the study of a small selection of the samples.[13] This team, consisting of scientists with differing backgrounds from MSC, USGS, and other government and university laboratories, was the first group to examine lunar samples before they were released to the sample PIs.

This final summary report and the four preceding reports were read with great interest by all of us at NASA headquarters. Each report contained some new and exciting revelation. LSAPT identified two types of rocks, crystalline and aggregates (later classified as breccias), as well as a variety of fine material from the lunar soil. Although the minerals in the rocks were similar to minerals found in the Earth's crust, there was a major difference. They contained a larger percentage of refractory elements such as titanium and zirconium. To a mineralogist this finding was important, leading LSAPT to proclaim that this mineral assemblage provided "difficulties for the fission hypothesis," that is, that the

Moon had been torn away from an early Earth by some cataclysmic event. If this had occurred, the minerals found in the samples should have been similar to those found on Earth.

Another, less hypothetical, conclusion was that the crystalline rocks were basalts, yet their density was greater than the average density of the Moon as a whole. This finding made it difficult to conclude that the Moon was a differentiated planetary body like the Earth, as it was thought to be, where the heavier material would be expected in the interior and the overall density of the planet should be higher than the density of rocks found at the surface. But this finding was consistent with the discovery of the "mascons," since this dense material was found in mare basins. If one pursued this logic, then some large portion of the Moon must be made up of less dense material to account for the difference, or else the Moon's core, if it had one, would have to be very small. It seemed clear that at some point part of the lunar surface had been molten.

After LSAPT performed its functions, it combined forces with the Lunar Sample Preliminary Examination Team to do more complete analyses and publish the results. To some degree this report skimmed the cream from the discoveries that would be announced later, but it served the important function of preparing us for the next missions. If we had had to wait for the sample PIs to report their findings we would have had little chance to modify or change the experiments and sampling procedures for *Apollo 12* and the later missions. The LSPET report, published in *Science* two months after *Apollo 11* returned, listed eighteen conclusions.[14] The most important from my perspective, paraphrasing the report's language, were that the crystalline rocks were different from any terrestrial rock and from meteorites; that the absence of hydrated minerals indicated there had been no surface water at Tranquility Base at any time since the rocks were exposed; that radioactive age dating showed they were crystallized 3 to 4 billion years ago; and that there was no evidence of biological material in the samples. Additional details and new findings would be released by the sample PIs four months later.

The Apollo 11 Lunar Science Conference was held in early January 1970 at the Rice Hotel in downtown Houston. The conference was an exciting time for all of us who had helped develop the Apollo science program. *Apollo 12* had returned to Earth just a little more than a month earlier, but all of its samples were still in quarantine and unstudied. Only the *Apollo 11* samples had undergone detailed examination by January 1970. Gary Latham, the principal investigator for the passive seismic experiment, had published a short report on his findings by this date along with the LSPET report mentioned above, but the sample PIs had agreed to withhold their findings until this meeting. Those performing the detailed sample analyses were all gathering at the same place for the first time. Approximately 1,100 PIs and their collaborators, including teams from sixteen foreign countries, had spent the past three months working feverishly to have their analyses ready for this day.

The expectation was palpable the first morning as we milled around in the hotel lobby. Whose theories would be confirmed, whose relegated to the dustbin of lunar science? Would any of the LSPET findings be challenged or changed?

Gene Shoemaker, representing his team from USGS and several universities, made the first presentation. He described the geologic setting of the lunar samples collected by the *Apollo 11* astronauts, coining the term "lunar regolith" for the surface characteristics at the landing site. The upper, regolith layer had been constantly churned and pulverized by impacts of all sizes. All the material returned had been collected from this fragmental debris layer, and the astronauts' observations had been made within 125 feet of the landing site. No "bedrock," or material in place, had been sampled. By geological standards it was not a very good collection of samples for such a large body as the Moon, but the consensus was that the samples were representative of a much larger area because of the mixing and transport of material brought in from afar as impact ejecta. Finally, he described the efforts to fix the location of each sample station. This had not been completely successful because the time limits for the EVA had restricted the number of photographs taken, but most had been located. Of the forty-seven pounds of material returned, approximately fifteen pounds had been distributed for analysis. (For the formal proceedings of the conference Shoemaker's presentation was modified and published as "The Apollo 11 Samples: Introduction.")[15]

Four days and 180 papers later the conference ended. We now had the first comprehensive view of one spot on the Moon based on data collected on the Moon itself. Several new minerals had been found, lunar lavas and breccias were common, and many samples bore evidence of shock metamorphism caused by impacts. *Science* devoted its entire January 30, 1970, issue to the conference. Though it is four times the size of a normal issue, it is a much more compact reference than the three-volume *Proceedings* for those who want to review the results of the first analyses of the *Apollo 11* samples in some detail.

The oldest samples dated gave radiogenic ages of approximately 4.7 to 4.9 billion years B.P. (before the present). Others gave dates of 4.13 to 4.22 and 3.78 billion years (some of the older dates were later disputed), in general much older than the first dates offered by LSPET. Only traces of carbon were found (one anomalous sample contained almost five hundred parts per million), and there was no evidence of any bio-organic compounds. One group of investigators (R. D. Johnson and C. C. Davis) stated that some of the high carbon readings might be attributable to contamination introduced during sample preparation or to errors in analytical techniques.[16] They suggested that an upper limit of ten parts per million would be correct for indigenous lunar organic material. They thought the small amounts of carbon detected in some of the samples might have come from the solar wind or from carbonaceous chondrites that had struck the Moon in ages past.

Water was not identified in any of the minerals analyzed, nor did Luis Alvarez find any magnetic monopoles. Some samples studied for remnant magnetism seemed to indicate that the Moon once had a small magnetic field, perhaps 1,000 to 1,500 gammas, or about one-thirtieth of the current field of the Earth. The present magnetic field was much smaller, however, on the order of 10 to 30 gammas, the latter figure coming from the magnetometer at the *Apollo 12* site that returned data by the time of the conference.

Preliminary results from measurements of the Laser Ranging Retro-Reflector were also reported. Accuracy in measuring the Earth-Moon distance had improved over that included in Mueller's report to the president four months earlier. This distance was now known to a precision of approximately one foot and was predicted to improve shortly to about six inches.

The Solar Wind Composition experiment carried on *Apollo 11* was not discussed at the conference. This experiment, mentioned in chapter 7, consisted of a sheet of aluminum foil hanging from a pole. After being exposed for seventy-seven minutes on the lunar surface, it was retrieved and brought back to Earth and placed in quarantine in case some lunar soil had adhered to the foil. When released from quarantine, it was carefully packed and sent to Switzerland for analysis by its PI, Johannes Geiss. He had made a quick analysis of the gases captured on the foil, finding noble gas ions as expected, and had reported his results in December in *Science*.[17] Eventually he extended his *Apollo 11* findings based on data returned from the next four missions, examination of pieces of *Surveyor 3* returned by *Apollo 12,* and data from the Vela satellites. Compiling all this information after his last experiment returned from the Moon, he stated in 1972 that he was now able to make good approximations of the average solar wind–noble gas abundances and ratios.[18] He forecast that a better understanding would evolve of the abundances of noble gases in the Sun and the atmospheres of Venus, Mars, and the major planets.

Latham's passive seismic experiment included in the Early Apollo Scientific Experiments Package continued to operate intermittently for twenty-one days. It survived the first lunar night before succumbing to the heat of the second day. Initial data telemetered to Earth had caused some consternation in Latham and the other members of his team. The Moon, based on these early data, seemed to be highly active seismically (apparently recording many small moonquakes), contrary to what had been predicted. After the first data had been analyzed, Ed Davin remembers walking between the Mission Control Center and the press conference room at MSC with Frank Press and Maurice Ewing, two of Latham's coinvestigators. They were trying to figure out what to tell the assembled reporters about this unexpectedly active Moon, apparently more active than the Earth. They asked Ed for his opinion, and he recalls being shocked that two of the world's leading seismic authorities would ask a lowly civil servant such a profound question. Ed could not suggest a solution, so Press and Ewing ended up announcing that the Moon appeared to be more active than the Earth, a new and disturbing "scientific discovery."

Eventually the explanation for this totally unexpected finding became clear. The lunar module landing stage, left behind when the astronauts departed, was creaking and groaning under the thermal stress of the wide temperature swings between lunar day and lunar night. In addition, the LM and backpacks discarded on the surface continued to emit gas long after the astronauts departed. Each quiver and burp of gas was being detected by the extremely sensitive seismometer just sixty feet away. These disturbances appeared in the data stream as small moonquakes. No one had anticipated that such tiny movements would be measured. Thus does science advance as we try to fit new data into old theories: some mysteries are quickly resolved.

The Moon, in fact, is seismically quiet (as opposed to Earth, where large or small earthquakes are being recorded almost constantly), and this was shown again and again as we deployed four more seismometers. Once the residual effects of the SIVB and LM impacts that occurred on later missions and the astronauts' presence had faded, the Moon stopped shaking. It was disturbed frequently by small movements believed to be caused by lunar tides (movements in the Moon's crust as a result of Earth-Moon interactions), thermal changes at sunrise and sunset, small impacts, or what were interpreted as rockfalls on nearby crater rims or mountainsides. A few larger true moonquakes were also recorded, with widely scattered epicenters concentrated at a depth of five hundred to six hundred miles, believed to be the base of the lunar mantle. The man-made shocks from the SIVB impacts also contributed to determining the thickness of the lunar crust.

Based on several years of data analysis, Latham and his team drew a number of conclusions. Below the thick lunar crust and mantle, constituting a "dynamically inactive outer shell," was a "core" with "markedly different elastic properties," and the core was very small. They believed that the core was at or near the melting point, but this did not "imply a major structural or compositional discontinuity as it does for Earth. However, the presence of a true core . . . is not precluded by present data." They also believed that "the presence of a thick lunar crust suggests early, intense heating of the outer shell of the Moon."[19] This last conclusion seemed to be validated by the visual evidence of widespread maria that filled all the low elevations on the Moon's nearside. Recent results from the Lunar Prospector mission appear to confirm Latham's findings and indicate that the Moon's core probably contains less than 4 percent of its mass, whereas the Earth's core makes up 30 percent of its total mass.

Continuing now from the findings above, where do we stand in answering the questions that had perplexed many noted scientists for centuries? Most students of the Moon would agree, I believe, that satisfactory answers are now in hand for most of those questions, although there is still no unanimous interpretation. Why should study of the Moon be different from other scientific controversies?

The burning question before the unmanned and manned missions—whether the craters observed were mostly of impact or volcanic origin—had been resolved to the satisfaction of most lunar students long before the first Apollo landing. Impacts were the answer, and Apollo data confirmed this conclusion. But the returned samples clearly showed that lava sheets or flows covered large areas of the Moon. What mechanism caused these flows is a little more debatable. Heating and melting of the lunar crust and mantle as a result of huge impacts is the favored explanation, not volcanic eruptions.

Next, where did the Moon come from? There is still some debate on this, but the possibilities have been narrowed and a preponderance of opinion favors one origin. Lunar samples show that the Moon's composition is similar to that of Earth, yet different. The Moon is not compositionally exotic, as proposed by Harold Urey and others, thus it probably was not captured early in Earth's history after having been formed somewhere else in the solar system. That leaves two theories: that it formed separately at about the same time as the Earth or that the Moon was split off from Earth by some event early in the Earth's formation.

Because the mineral assemblages found in lunar samples differ somewhat from rocks that have formed on Earth, either origin is possible. However, the Moon most likely was torn from the Earth by the impact of another large body that contributed some of its material to the Moon, thus accounting for the mineralogical differences. This latter theory is gaining more and more favor in recent years as other conditions, such as the Moon's angular momentum, become better understood and are factored into the models being used.

The next question, How old is the Moon? can now be answered with some certainty. Age dating of lunar samples has shown extremely old ages, some as high as 4.4 to 4.5 billion years B.P. This rivals the oldest ages found in meteorites, which until this point were the most ancient objects dated. This date agrees with the thinking of most solar system students about when the solar nebula began to clump and form the planets, indicating that the Moon formed almost simultaneously with the Earth at a very early point in the birth of our solar system. The "genesis rock," collected on *Apollo 15*, is almost pure anorthosite, a type of rock formed on Earth at great depths. It is believed to represent a piece of the Moon's early crust. Argon-argon dating found an age of crystallization of approximately 4.0 billion years B.P.[20] However, this type of dating can produce lower than actual ages; thus the "genesis rock" may be older—closer to 4.4 to 4.5 billion years.

Whether there has ever been water on the Moon, or whether water still exists there, has been a continuing and intriguing question. None of the samples analyzed showed that water was present during the formation of the lunar crust. But in March 1971 John Freeman of Rice University, the PI for the Suprathermal Ion Detector Experiment (SIDE), reported that he had recorded the occurrence of water vapor for three "events" at his instruments left at the *Apollo 12* and *Apollo 14* sites. These measurements had been made at the same time Gary Latham recorded a swarm of moonquakes, suggesting that the two events were connected. Earlier Freeman had recorded the LM and SIVB impacts as disturbances in the Moon's ionosphere, but these events had a different character than those he believed indicated water vapor. This created a stir in the media that prompted us to try to put Freeman's claim in a larger context.[21]

Acknowledging the importance of discovering water on the Moon, we discussed potential sources of the inferred water vapor, possibly related to material left behind by the astronauts in the LM descent stage tanks, portable life-support system tanks, and other items discarded on the lunar surface. We also pointed out that the SIDE experiment identified the mass of ions (in a gas cloud) only in a range of energy that would also include methane or neon, which could also have a lunar origin. Ultimately Freeman's recordings were not considered conclusive in detecting water.

The recent lunar probe, Lunar Prospector, appears to support the possibility that water, in the form of ice, exists on the Moon in the permanently shadowed craters near the poles. If ice is present, it is most probably a by-product of comet impacts. Sensors on Lunar Prospector detected hydrogen, and the most likely source of the hydrogen is considered to be ice. Perhaps Freeman had detected an early whiff of water vapor from his two experiments.

To sum up the operational accomplishments of the six Apollo landing missions: almost 5,000 pounds of experimental equipment were landed on the Moon, and 840 pounds of lunar material (rocks, dirt, drill cores, etc.) were returned under carefully controlled conditions. Five ALSEPs, which included most of the total of fifty-three individual experiments deployed by the astronauts while on the lunar surface, were placed at different locations. And approximately sixty miles of traverses were recorded, both on foot and using the LRV, in support of the field geology studies and geophysical surveys. In addition, detailed data were collected on missions 15, 16, and 17 from instruments carried in the command and service module, including photographs, compositional analysis of broad areas of the Moon's surface, mapping its magnetic and gravity fields, and analyzing its tenuous atmosphere. All of these data contributed toward deciphering the Moon's many mysteries as well as resolving less controversial issues.

For young engineers dreaming of one day building lunar bases, the Moon will be a friendly place. Lunar bulldozers and backhoes will be able to excavate and move lunar soil just as we move soil on Earth. There will be obvious differences, but we gained sufficient data through the soil mechanics experiment and other experiments to design such machines. Structures could be covered with lunar soil to shield them from solar flares and high energy particles, thus obviating the need to bring shielding from Earth. If needed, "regolith blocks" could be made from the soil that would be as useful as terrestrial cinder blocks. Unlike bases built in Antarctica, the closest terrestrial analogue to lunar bases, which must be constantly refurbished or rebuilt because of damage from snow and ice, lunar bases once constructed should last for the ages. Only a direct hit or near miss from a meteorite could damage the base. And perhaps if bases are built near the Moon's poles the Moon can be mined for water, the most valuable of all lunar resources. The Apollo program provided the shoulders to stand on—now it is up to future explorers to go beyond our "giant leap for mankind."

A few more words concerning the results of the Clementine and Lunar Prospector programs. Both of these programs continue to add to our knowledge of the Moon. In some instances they are expanding on what we learned from Apollo, and in other exciting ways they are providing new information. Rather than my attempting to summarize their results to date, references in the notes discuss some of the findings.[22] Many other papers and reports discuss the results of these two missions.

The final maneuver for Lunar Prospector, a last-minute addition to its scientific objectives, was a controlled crash similar to those carried out by the Apollo LM ascent stages and SIVBs. This time the impact point selected was a perpetually shadowed crater near the Moon's south pole, in the hopes that telescopes in orbit or on Earth would record the plume from the crash and confirm the presence of water. Such a cloud was not seen, repeating our experience during the Apollo missions when I asked observatories in France with large telescopes to try to observe and measure the impact of the *Apollo 16* SIVB stage. This would have been a much larger event than was expected for the Lunar Prospector impact. The time of the *Apollo 16* SIVB impact prevented any United States observatories from participating, since the Moon would be below the horizon. The weather was not completely cooperative when observatories at Meudon, Pic-du-Midi, and Nice attempted to observe the impact on the night of April 19, 1972, and this might have accounted for the negative report we received.[23] However, the failure to see a cloud at the impact point selected for Lunar Prospector's final act will not detract from its successes; further analysis of data recorded by the spacecraft's sensors will without doubt continue to add to our understanding of the Moon in the years ahead.

In successfully undertaking the challenge set by President Kennedy (with emphasis on "successfully"), Apollo taught us one final lesson. Apollo's heritage went far beyond knowledge about the Moon and Earth. Now that many of the records of the former Soviet Union have been opened to public scrutiny, it has been confirmed that we really were in a race to the Moon. It certainly seemed that way to us at the time, but you could not be sure because Soviet launches were always veiled in secrecy; the world became aware of them only after they were on their way to whatever destination, and failures were never reported. The Soviets' long-range plans were seldom discussed, although Boris Voishol, from the Soviet Tectonic Academy, writing in the September 1968 *Geotimes,* stated: "The first landing of Soviet cosmonauts on our moon is scheduled in the near future."[24]

Based on information available at that time, the missing ingredient in their ability to send men to the Moon was a booster as large as the *Saturn V,* which would be needed for the round trip. Without such a rocket we assumed that if they were really intent on a manned lunar landing they would use their smaller, proven rockets to assemble the needed launch capability in Earth orbit before going on to the Moon—one of NASA's original proposals. We now know that they were building a *Saturn V*–class rocket but that on its first test flight it crashed shortly after lift-off. On a second launch attempt a few months later, it exploded on the pad, apparently killing some of their rocket experts, and was never rebuilt.[25] The Soviet failures—and there were many—were only a matter of speculation for most of us, though undoubtedly there were some who were privy to intelligence sources and knew about their difficulties. Our launches, successful or unsuccessful, were always made in full view of the world.

What if the Soviet Union had landed men on the Moon first? Several writers have discussed the effect of Project Apollo on the Soviet Union; here is another view with which you may or may not agree. My father, a civilian stationed in West Germany for the Army Signal Corps at the time of the launch of *Sputnik I,* remembered an unnerving encounter with one of his German contractors. When it was confirmed that the Soviets had successfully orbited the first satel- lite, this man came running up with fear on his face. His conclusion was that this demonstration of Soviet technological superiority spelled doom for the world. The United States failures at launching the Vanguard rocket were well known. Suddenly the Soviet Union had leapfrogged our efforts. Along with its newly demonstrated nuclear weapons, this made the man believe the bad guys had won the Cold War. We would soon have to knuckle under to this new dominant world force. He was seeking reassurance that his analysis was wrong, but with limited knowledge of how our space programs were proceeding, my father could not give it.

The point of this anecdote is to show how fragile a nation's leadership is in a rapidly evolving world. In view of their recent history, West Germans in October 1957 might be forgiven for being pessimistic. But as I remember, this pessimism was widespread even in the United States, with finger pointing and blame all around for our inability to beat the Soviets during the early days of space flight.

What would the world look like today if the Soviets' program had not experienced its hidden failures and they had been first to land men on the Moon? I suspect it would be different, but of course there is no way to prove it. Everyone likes a winner and gravitates toward one regardless of worthiness; second place seldom attracts much enthusiasm. Accommodation to Soviet leadership would have been rationalized, and the Soviet bloc might have become the dominant force in world politics, perhaps postponing or averting its ultimate economic collapse. Meanwhile, we would be scrambling to catch up and demonstrate that a democratic government could do as well as or better than a state-directed totalitarian government. Remember, in the 1960s many countries were experimenting with or embracing communist forms of government, and Soviet-led expansion of communist ideology was making great strides even without demonstrating the overall technological leadership that a "first" on the Moon would have given.

What is the lesson of Apollo that goes beyond being the first to land on the Moon and the expansion of our scientific knowledge? It seems pretty basic. Free societies can successfully undertake enormously complex actions—if they dare. Although the United States was the leader in Apollo, many other nations contributed people, technology, or facilities. Apollo was a dream that everyone could embrace, if permitted, and all could share in the sweet reward of success. The few words on the plaque carried by *Apollo 11* said it all: it was an accomplishment "for all mankind."

Conceived primarily as a political statement, Apollo achieved much more than its original goal. Now, when faced with seemingly intractable problems, someone will be heard to say, "If we can land a man on the Moon, why can't we [fill in the appropriate objective]?" And of course that is the right question to ask, because people of goodwill, working together, are capable of solving very difficult problems. Apollo proved it. Let's not forget that dreaming big has its own rewards, even if occasionally we stub a toe. That is the essential lesson I carry away from my Apollo days, and I hope it will be remembered by those who study and follow our example in the future.

Notes

Abbreviations used in the notes that are not in the list of abbreviations and acronyms are internal NASA office codes, used for mail purposes.

I used these four sources to verify numbers and recall details of the missions:

Preliminary Science Reports for Apollo 11, SP-214; *Apollo 12,* SP-235; *Apollo 14,* SP-272; *Apollo 15,* SP-289; *Apollo 16,* SP-315; and *Apollo 17,* SP-330 (Washington, D.C.: NASA, Scientific and Technical Information Branch, 1969–73).

Three NASA guidebooks by Gene Simmons, chief scientist, Manned Spacecraft Center: "On the Moon with *Apollo 15,*" June 1971; "On the Moon with *Apollo 16,*" April 1972; and "On the Moon with *Apollo 17,*" December 1972.

NASA press kits for *Apollos 15, 16,* and *17.*

Apollo Scientific Experiments Data Handbook, NASA TM X-58131, W. W. Lauderdale, General Electric Co., and W. F. Eichelman, JSC, technical editors (Washington, D.C.: NASA, August 1974).

1 From the Jungle to Washington

1. "Report of the Ad Hoc Working Group on Apollo Experiments and Training on the Scientific Aspects of the Apollo Program," internal NASA document, December 15, 1963. The draft of the same report was dated July 6, 1962.

2. *A Review of Space Research,* Publication 1079 (Washington, D.C.: National Academy of Sciences–National Research Council, 1962).

3. Memo, From: SM/Acting Director, Manned Space Sciences Division Office of Space Sciences, To: Director, Manned Spacecraft Center, Attention: Robert R. Gilruth, Subject: Scientific Guidelines for the Apollo Project, October 8, 1963.

4. C. A. Pearse and H. W. Radin, *Lunar Logistic System Scientific Facility* (Washington, D.C.: Bellcomm, January 31, 1963).

5. Robert F. Fudali et al., *An Analysis of the Value of a Lunar Logistic System,* part 2, *The Scientific Exploitation of the Moon* (Washington, D.C.: Bellcomm, March 18, 1963).

6. *Lunar Logistic System,* 10 vols., MTP-M-63-1 (Huntsville, Ala.: Marshall Space Flight Center, March 15, 1963).

2   Early Theories and Questions about the Moon

1. Ralph B. Baldwin, *The Measure of the Moon* (Chicago: University of Chicago Press, 1963).

2. Zdenek Kopal and Zdenka Mikhalov, *I.A.U. Symposium on the Moon* (New York: Academic Press, 1962); Gilbert Fielder, *Structure of the Moon's Surface* (New York: Pergamon Press, 1961); Harold Clayton Urey, *The Planets, Their Origin and Development* (New Haven: Yale University Press, 1952); Urey, *The Origin and Nature of the Moon,* Annual Report (Washington, D.C.: Smithsonian Institution, 1960), 251–65; Gerard P. Kuiper, "On the Origin of Lunar Surface Features," *Proceedings of the National Academy of Sciences* 40 (1954): 1096; and L. W. LeRoy, "Lunar Features and Lunar Problems," *Geological Society of America, Bulletin* 72, 4 (1961): 591–604. In general terms, the texts listed above sum up the state of knowledge of the Moon at the beginning of Project Apollo, although many more contemporaneous references also dealt with lunar problems. Aside from anecdotal stories, I have depended primarily on these authors and a few other sources for the information contained in this chapter; I claim no original data, only a selective winnowing.

3. This mystery was partly solved in 1959 by a Soviet camera-carrying spacecraft, *Lunik 3;* although the photos had poor resolution, the farside appeared similar to the nearside highlands—highly cratered, but with no major maria, a nearside feature.

4. Ian Jackson, ed., *The Earth's Mantle: Composition, Structure, and Evolution* (New York: Cambridge University Press, 1998).

5. Letter to Dr. Jay Holmes, Chief Special Services, Manned Space Flight, NASA, Washington, D.C., January 10, 1964, signed by Harold C. Urey, a response to Holmes's letter discussing authors' theories on the origin of the Moon.

6. Thomas Gold, "The Lunar Surface," *Monthly Notices of the Royal Astronomical Society* 115 (1955): 585.

7. J. V. Evans and T. Hagfors, "On the Interpretation of Radar Reflections from the Moon," *Icarus* 3 (1964): 151.

8. Bruce W. Hapke and H. Van Horn, "Photometric Studies of Complex Surfaces with Applications to the Moon," *Journal of Geophysical Research* 68 (1963): 4545.

9. Grove K. Gilbert, "The Moon's Face," *Bulletin of the Philosophical Society of Washington,* 1893.

10. Eugene M. Shoemaker, "Impact Mechanics at Meteor Crater, Arizona," in *The Solar System,* vol. 4, *The Moon, Meteorites, and Comets,* ed. Barbara M. Middlehurst and Gerard P. Kuiper, 301–36 (Chicago: University of Chicago Press, 1963).

11. Robert F. Fudali, Brief résumé of trip to Menlo Park and Ames Research Center, memo for file, Bellcomm, Washington, D.C., April 20, 1963.

12. "Lunar Scientific Model," project document 54, Jet Propulsion Laboratory, Caltech, Pasadena, Calif., December 15, 1966.

13. K. Watson, B. Murray, and H. Brown, "On the Possible Presence of Ice on the Moon," Letters to the Editor, *Journal of Geophysical Research* 66, 5 (1961): 1598–1600.

14. For a further discussion of the major lunar scientific questions posed before the Apollo missions began, see the 1966 report of the National Academy of Sciences, *Space Research: Directions for the Future,* National Academy of Sciences–National Research Council Publication 1403 (Washington, D.C.: National Academy of Sciences, 1966).

15. Baldwin, *Measure of the Moon,* xviii.

16. James C. Greenacre, "A Recent Observation of Lunar Color Phenomena." Apparently I obtained a typed transcript from an undated tape recording that Greenacre made immediately following his observations. A similar report was later published in *Sky and Telescope* 26, 6 (1963).

17. John S. Hall, "Supplementary Report," Lowell Observatory, November 5, 1963, unpublished copy.

18. "Lunar Color Phenomena," ACIC Technical Paper 12, May 1964, Prepared by Lunar and Planetary Branch, Cartography Division. (It has good drawings of the feature.)

19. B. M. Middlehurst, J. M. Burley, P. A. Moore, and B. L. Weither, "Chronological Catalog of Reported Lunar Events," Interim Draft, G-807, NASA, probably circulated in late 1967. A final report may have been written.

20. NASA memo, To: MCL/J. Holmes, From: MAL/D. Beattie, Subject: Tantalizing Tidbit for Dr. Mueller's Senate Hearing, Date: April 28, 1969. My memo does not provide a reference to the report I was quoting from.

21. W. C. Cameron, "An Appeal for Observations of the Moon," *Journal of the Royal Astronomical Society of Canada* 59 (1965): 219; Cameron, "Observations of Changes on the Moon," in *Proceedings of the Working Group on Extraterrestrial Resources* 47-56, Fifth Annual Meeting, March 1–3, 1967; Cameron, "Lunar Transient Phenomena (LTP): Manifestations, Site Distribution, Correlations and Possible Causes," *Physics of the Earth and Planetary Interiors* 14 (1977): 194–216.

3   What Do We Do after Apollo?

1. Robert C. Seamans Jr., *Aiming at Targets: The Autobiography of Robert C. Seamans, Jr.* NASA History Series, SP-4106 (Washington, D.C.: NASA, Scientific and Technical Information Division, 1996).

2. NASA memo, To: M/Associate Administrator, From: S/Associate Administrator, Subject: Modifications to LEM Periscope, date missing.

3. *Lunar Logistic System,* vol. 10, *Payloads,* MTP-M-63-1 (Huntsville, Ala.: Marshall Space Flight Center, 1963).

4. Memo from the National Academy of Sciences, National Research Council, for Dr. H. E. Newell, Associate Administrator, Space Science and Applications, Subject: Future

Goals of the Space Science Program, Date: August 11, 1964, signed by H. H. Hess, Chairman, Space Science Board.

5. *Washington Daily News,* January 27, 1965; headline reads: "LBJ Says U.S. Intends 'to Explore the Moon, Not Just Visit or Photograph It.'"

6. *Lunar Logistic System,* vol. 9, *Mobility on the Lunar Surface* (Huntsville, Ala.: Marshall Space Flight Center, 1963).

7. Jean R. Olivier and David C. Cramblit, Lunar Surface Mobility Study, Internal Note, LTIR-2-DF-62-5, December 14, 1962, NASA Launch Operations Center, Future Studies; Jean R. Olivier and Richard E. Valentine, Engineering Lunar Surface Model Obstacles (ELMO), TR-145-D, John F. Kennedy Space Center, March 8, 1965.

8. "Surveyor Project, Lunar Scientific Model," Project Document 54, Jet Propulsion Laboratory, California Institute of Technology, Pasadena, Calif., December 15, 1966.

9. "Bendix Selects Four-Wheeled Vehicle for Mobile Lunar Lab," *Aviation Week and Space Technology,* December 7, 1964.

10. E. H. Wells, "Optical Astronomy Package Feasibility Study for Apollo Applications Program," Executive Summary Report, TM X-53496, Marshall Space Flight Center, August 5, 1966.

4 The United States Geological Survey Joins Our Team

1. Don E. Wilhelms, *To a Rocky Moon: A Geologist's History of Lunar Exploration* (Tucson: University of Arizona Press, 1993).

2. NASA memo, To: NASA Headquarters, Attention: Dr. Verne C. Fryklund Jr., SM, From: Assistant Director for Engineering and Development, Subject: Vacuum Chambers and Solar Simulators, Date: June 19, 1964.

3. U.S. government memo, To: MLEI Personnel, From: Gordon A. Swann and J. T. O'Connor, Subject: Use of Apollo-Type Space Suit at MSC, Date: June 23, 1965.

4. J. D. Friedman, R. J. P. Lyon, D. A. Beattie, and J. Downey, "Lunar Ground Data Required for Interpretation of AES Orbital Experiments, Post-Apollo Space Exploration," *American Astronautical Society, Advances in the Astronautical Sciences* 20 (May 1965).

5. "NASA Summer Conference on Lunar Exploration and Science," NASA SP-88, Falmouth, Mass., July 19–31, 1965.

6. Gerald G. Schaber, Technical Letter: "Astrogeology—11, Apollo Applications Program Investigations, Field Test 3," U.S. Department of the Interior, Geological Survey, prepared for the National Aeronautics and Space Administration, undated.

7. Letter to Administrator Thomas O. Paine from USGS Director W. T. Pecora, June 29, 1970.

8. "Opportunities for Participation in Space Flight Investigations," NHB 8030.1A, April 1967, NASA, Washington, D.C.

9. During the first week of the conference the astronomy working group met at the University of Virginia, chaired by our old friend Larry Fredrick. The results of this

meeting were reported to the conference in Santa Cruz by Nancy Roman, a NASA headquarters astronomer.

10. "1967 Summer Study of Lunar Science and Exploration," NASA SP-157, University of California–Santa Cruz, July 31–August 13, 1967.

5 Science Payloads for Apollo: The Struggle Begins

1. NASA memo, To: SM/Dr. V. C. Fryklund, From: MGL/D. A. Beattie, Subject: Suggested Priority List for Twenty-four-Hour Apollo Experiment Categories, Date: November 21, 1963.

2. NASA memo, To: SSSC Chairman, From: SM/Director, Subject: Preliminary Definition of Apollo Investigations, Date: February 13, 1964.

3. NASA memo, To: SM/Director, Manned Space Science Division, SM/Chief, Lunar and Planetary Science Branch, From: MTF/D. A. Beattie, SM/E. M. Davin, SM/P. D. Lowman, Subject: Apollo Scientific Investigations, Date: June 12, 1964.

4. NASA memo, To: SM/Director, Manned Space Science Division, SM/Chief, Lunar and Planetary Science Branch, From: MTF/D. A. Beattie, SM/E. M. Davin, SM/P. D. Lowman, Subject: Supplemental Recommendations and Comments on "Apollo Scientific Investigations" (Davin, Beattie, and Lowman, 6-12-64), Date: July 6, 1964.

5. Letter from the National Academy of Sciences, Space Science Board, To: Dr. Homer E. Newell, Associate Administrator, Office of Space Science and Applications, Subject: Report of Ad Hoc Committee on Lunar Sample Handling Facility, Date: February 2, 1965, signed by Harry H. Hess, Chairman.

6. *Apollo Program Development Plan,* NPC C500, MA 001.000 (Washington, D.C.: Apollo Program Office, Office of Manned Space Flight, NASA, 1965).

7. R. F. Fudali, Trip Report, Subject: Planetology Subcommittee of the Space Science Steering Committee Meeting at UCLA, January 21–23, 1965, Date: January 27, 1965.

8. "Technical Plan for the Scientific Exploration and Utilization of the Moon," by Manned Space Science Program, OSSA, NASA Headquarters, Washington, D.C., June 1965.

9. *Apollo Experiments Guide,* NPC 500-9 (Washington, D.C.: NASA Office of Manned Space Flight, 1965).

10. "Apollo In-Flight Experiment Guide, Preliminary," June 11, 1964. (It is not clear which office in OMSF issued this guide, but it may have been the Apollo Flight Operations Division [Code MAO], since potential experimenters were directed to that office to obtain the forms needed.)

11. NASA memo, To: AA/Associate Administrator, From: M/Associate Administrator for Manned Space Flight, S/Associate Administrator for Space Science and Applications, Subject: "Lunar Exploration Plan," September 24, 1965, and revisions.

12. "Flight Control Handbook for Experimenters," Flight Operations Directorate, MSC, Houston, Texas, April 13, 1967.

13. One court reporter would type the conversations on a steno machine, and a

second would transcribe the tape, which could then be called up on your console if you wanted to check what had been said. The complete rough transcripts were also available between EVAs to help plan the next EVA. Gordon Swann estimates that the time lapse between the spoken word and the tape's being made available was thirty to sixty seconds; thus it was a valuable resource that was used extensively by the team in the SSR.

14. The actual procedure was a little more complicated. SSR questions were passed up on the voice link to someone like James Lovell, who screened them and passed them on to the flight director, then they went to the CapCom. With all this review, the question that finally arrived at the CapCom console was often very different from the original.

15. NASA memo, To: Distribution, From: MAL/Beattie, Subject: MCC-H Simulations for Apollo 14, Date: September 24, 1970.

16. NASA memo, To: MA/Dr. J. Turnock, From: MTL/D. Beattie, Subject: Trip Report—MSC, September 13, 14, 15, and 16, Date: September 18, 1967.

17. Courtney G. Brooks, James M. Grimwood, and Loyd S. Swenson Jr., *Chariots for Apollo: A History of Manned Lunar Spacecraft*, NASA History Series SP-4205 (Washington, D.C.: NASA, Scientific and Technical Information Branch, 1979), 201–2.

18. NASA memo, To: MAL/Assistant Director for Lunar Science, From: MAL/Program Manager, Plans and Objectives, Subject: Trip Report—Lunar Surface Operations Planning Meeting, MSC, Date: February 16, 1968.

19. NASA memo, To: MT Staff, From: MTD/Deputy Director, Advanced Manned Missions Program, Subject: Reorganization of the Advanced Manned Missions Program, Date: December 8, 1967.

20. NASA Special Announcement, Subject: Establishment of an Apollo Lunar Exploration Organization within OMSF, Date: December 19, 1967, signed by John E. Naugle, Associate Administrator for Space Science and Applications, and George E. Mueller, Associate Administrator for Manned Space Flight.

21. Letter to Mr. James E. Webb, Administrator, National Aeronautics and Space Administration, Washington, D.C., Date: April 2, 1968, signed by Charles H. Townes, Chairman for the Scientific and Technology Advisory Committee.

22. N. W. Hinners, D. B. James, and F. N. Schmidt, "A Lunar Exploration Program," Technical Memorandum, TM-68-1012-1, Bellcomm, Inc., January 5, 1968.

23. NASA memo, To: NASA Headquarters, Attention: Captain Lee R. Scherer, From: Director of Lunar Exploration Working Group, Subject: Comments on the Proposed Lunar Exploration Memorandum, July 31, 1968, signed by John D. Hodge.

24. NASA memo, To: Distribution, From: Lunar Exploration Working Group, Subject: Lunar Exploration Program Memorandum, Date: October 4, 1968, cover memo signed by Lee R. Scherer, Director, Apollo Lunar Exploration Office.

25. D. A. Beattie and F. El-Baz, "Apollo Landing Site Selection," *Military Engineer* 62, 410 (1970): 370–76.

26. NASA memo: To: MAL/Distribution, From: MAL Director, Subject: MAL FY 1969 Operating Plan, Date: November 19, 1969; the memo noted our office's contribution of $700,000.

6 Developing the Geological Equipment, Experiments, and Sampling Protocols

1. NASA memo, From: SM/Director, Manned Space Science Division, To: SSSC/Chairman, Space Science Steering Committee, Subject: Preliminary Definition of Apollo Investigations, Date: February 13, 1964.

2. Problems encountered by the astronauts on *Apollo 14* and discussed in chapters 8 and 11 show that these concerns were well founded.

3. Letter to Conrad Boette, Manned Spacecraft Center, Houston, Texas, and John W. Hardin, Marshall Space Flight Center, Huntsville, Alabama, from Gordon Swann, Branch of Astrogeology, Flagstaff, Arizona, Re: Lunar Periscope, Date: March 1, 1965.

4. NASA memo: To: Director, Manned Spacecraft Center, Attention: E/Assistant Director for Engineering and Development, From: SM/Director, Manned Space Science Programs, Subject: Proposed Guidelines for Manned Lunar Science, Date: September 13, 1965, signed by Willis B. Foster.

5. Additional proposals were approved in 1967 by a committee chaired by Jim Arnold of the University of California, San Diego; they added 101 more PIs, making a grand total of 142 teams that would study the first samples brought back from the Moon.

6. Bellcomm memo for file, From: B. E. Sabels, Subject: Apollo Lunar Hand Tools Critical Design Review, MSC, February 23-24, 1967, case 340, Date: March 2, 1967.

7. NASA memo, To: MA/James H. Turnock, From: MTL/Donald A. Beattie, Subject: Description of Apollo "grab sample," Date: September 28, 1967.

8. Bellcomm memo for file, From: B. E. Sabels, Subject: Apollo Lunar Sample Return Container, Final Design Review, Oak Ridge, Tennessee, February 9–10, 1967, case 340, Date: February 21, 1967.

9. Oral Presentation on Contract NAS 9-4860, Investigation of Lunar Surface Contamination by LEM Descent Engine and Associated Equipment, Grumman Aircraft Engineering Corporation, Arthur D. Little, Inc., November 2, 1965.

10. Paul Lowman was the scientist responsible for interpreting this vast catalog of new data, and he published striking sets of terrain and weather photographs taken during the Gemini and Apollo Earth orbital missions along with an explanation of what the pictures told us. Paul D. Lowman, *Space Panorama* (Zurich: Weltflugbild Reinhold A. Müller, 1968), and Lowman, *The Third Planet* (Zurich: Weltflugbild Reinhold A. Müller, 1972).

11. NASA memo, To: M/Associate Administrator for Manned Space Flight, From: S/Associate Administrator for Space Science and Applications, Subject: Requirements for Lunar Surface Stereo Photography, Date: November 23, 1966.

7 The Apollo Lunar Surface Experiments Package and Associated Experiments

1. NASA memo, From: SM/Director, Manned Space Science Division, To: SSSC/Chairman, Space Science Steering Committee, Subject: Preliminary Definition of Apollo Investigations, Date: February 13, 1964.

2. NASA memo, From: Director, Research Projects Laboratory, R-RP-DIR, To; Dr. Wernher von Braun, DIR, Subject: Apollo Scientific Station Assignment, Date: May 13, 1965.

3. NASA memo, From: MTF/Donald A. Beattie, To: MTG/Director, Manned Planetary Mission Studies, Subject: Saturday Meeting with Dr. Mueller, Date: May 24, 1965.

4. Bendix Systems Division Press Release, From: Daniel H. Schurz, Public Relations Office, For Immediate Release (No Date), Three Firms Selected to Design Apollo Lunar Surface Package; "NASA Selects Lunar Package Design Firms," *Marshall Star*, August 18, 1965, 3.

5. NASA memo, To: C. J. Weatherred, From: Program Manager, EASEP (No subject, but purpose of memo was to commend Bendix for work on EASEP), Date: September 22, 1969.

6. Draft attachment that I wrote to what I presume was to be a letter from Petrone to Gilruth on surface experiments problems and concerns, September 3, 1970.

7. "Research News," *Science* 177 (August 11, 1972): 506.

8. J. R. Bates, W. W. Lauderdale, and H. Kernaghan, "ALSEP Termination Report," NASA Reference Publication 1036, April 1979.

9. Meher Antia, "Neutron Stars Spin out Gravity Waves," *Science* 280 (June 19, 1998): 1835.

10. Conversations with Robert Grilli, James Bates, and Charles Capps of Johnson Space Center. Some of the stories in this chapter are based on conversations with Richard J. Green, Edward M. Davin, Eugene Zaitzeff, and Charles J. Weatherred.

8 Walk, Fly, or Drive?

1. NASA memo, To: Distribution, From: MAL/D. A. Beattie, Subject: Trip Report, April 9, 1968. Report on meetings at MSC to develop the Announcement of Flight Opportunities for CSM orbital science, work on the Advanced ALSEP, and suit development.

2. NASA letter from Donald A. Beattie to James V. Correale, Assistant Chief for Test and Development, Crew Systems Division, Code EC, MSC, Houston, Texas, 77058, discussing EVA activities to be considered for lunar EVA suits "for 1971 missions and beyond," April 10, 1968.

3. NASA news press kit, Project: Apollo 15, Release No: 71-119K, For release: Thursday A.M. July 15, 1971, Date: June 30, 1971. (Text details are from the press kit.)

4. NASA memo, From: MAT-2/A. F. Phillips, To: MAT-2/Chief, Apollo Spacecraft Test, Subject: Lunar Surface Activity—Some Rough Correlations with Earth Activity, Date: February 19, 1969.

5. NASA memo, To: NASA Headquarters, Attention: Lt. Gen. Samuel C. Phillips, From: Director, Manned Spacecraft Center, Subject: Improvement of Apollo Space Suit for Lunar Surface EVA Tasks, Discussing Response to Phillips May 24, 1968 Letter Requesting Suit Improvements, Date: October 1, 1968, signed by George Trimble (?) acting for Robert R. Gilruth.

6. Memo, To: U. Liddel, From: V. Bremenkamp, Subject: LPMB Action on Lunar Exploration, with Attachment, Date: 15 April 1969. The memo originated from the Office of the President, Associated Universities, 1717 Massachusetts Avenue NW, Washington, D.C. 20036. (Associated Universities provided support to the Lunar and Planetary Mission Board at this time.)

7. Personal communication, Saverio "Sonny" Morea.

8. NASA memo, To: Marshall Space Flight Center, Manned Spacecraft Center, Kennedy Space Center, From: MA/B. Milwitzky, Subject: Lunar Roving Vehicle Requirements for Apollo Program Specification, Date: January 16, 1970.

9. Personal communication, Sonny Morea.

9 Astronaut Training and Mission Simulation

1. Flight Crew Training Plan, NASA General Working Paper 10,022, NASA, MSC, January 17, 1964.

2. R. F. Fudali and N. W. Hinners, memo for file: Trip Report—Prototype Apollo Space Suit Demonstration, MSC, September 20, 1963, case 110, Date: October 1, 1963.

3. R. F. Fudali, memo for file (draft): Subject: Test of the Apollo Spacesuit on Volcanic Terrain, case 211, September 9, 1964.

4. NASA memo, To: M/Associate Administrator for Manned Space Flight, From: SM/Director, Manned Space Science Division, Subject: Apollo Field Simulation, Date: September 8, 1964; NASA memo, To: M/Associate Administrator for Manned Space Flight, From: MT/Director for Advanced Manned Missions, Subject: Utilization of USGS Field Simulation Support, Date: September 25, 1964.

5. NASA memo, To: M/Associate Administrator for Manned Space Flight, From: MM/Deputy Director, Space Medicine, Subject: Apollo Field Simulation, Date: September 15, 1964.

6. Letter to George E. Mueller, Associate Administrator for Manned Space Flight, NASA, Washington, D.C. 20546, responding to copy of Foster's memo of September 8, 1964, dated September 25, 1964, signed by George M. Low, Deputy Director, MSC.

7. NASA memo, To: M/Associate Administrator for Manned Space Flight, From: SM/Director Manned Space Science Division, Subject: Lunar Simulation Facilities, Date: October 7, 1964.

8. Letter to Dr. Robert Gilruth, Director, Manned Spacecraft Center, NASA, Houston, Texas, Subject: Lunar Surface Mission Simulation, Date: October 21, 1964, signed by George E. Mueller.

9. D. A. Beattie and P. D. Lowman, "Origin of Laguna de Guatavita, Colombia," Abstracts of Programs, Geological Society of America Annual Meeting, November 1965.

10. Donald K. Slayton, Science and Technology Advisory Committee Briefing, Lunar Landing Mission, part 3, MSC, Crew Training and Simulation, October 6–8, 1967.

11. Trip report, John Hanley, Report on Grover Testing, August 31–September 4, 1970, Objectives: (1) Driver Training, (2) Hand Tool Stowage, undated.

10 Studying the Moon from Orbit

1. P. C. Badgley and R. J. P. Lyon, "Lunar Exploration from Orbital Altitudes," paper presented at New York Academy of Sciences Conference on Geological Problems in Lunar Research, May 16–19, 1964, New York. See also P. C. Badgley, "The Application of Remote Sensing in Planetary Exploration," paper presented at Third Annual Symposium on Remote Sensing of Environment, October 14, 1964, Ann Arbor, Michigan, and P. C. Badgley, W. A. Fischer, and R. J. P. Lyon, "Geologic Exploration from Orbital Altitudes," *Geotimes* 10, 2 (1965): 11–14.

2. NASA memo, To: AA/Robert R. Gilruth, From: MA/Major General Samuel C. Phillips, Subject: Lunar Photography from the CSM, Date: March 29, 1968.

3. NASA memo, To; Manned Spacecraft Center, Attention: Robert R. Gilruth, From: Apollo Program Director, Subject: Lunar Scientific Investigations from the CSM, Date: May 21, 1968.

4. Jack Trombka, personal communication.

5. The photographic equipment used for this study is described in some detail in *Preliminary Science Report for Apollo 12*, SP-235 (Washington, D.C.: NASA, Scientific and Technical Information Branch, 1970).

6. A libration point is any position in the plane of a celestial system consisting of massive bodies orbiting each other at which the gravitational influences of the bodies are approximately equal, creating in essence a region of zero g.

7. Details on these experiments can be found in *Preliminary Science Report for Apollo 14*, SP-272 (Washington, D.C.: NASA, Scientific and Technical Information Branch, 1971).

8. For more about these important experiments, see Gene Simmons's two NASA guidebooks, "On the Moon with *Apollo 16*," April 1972, and "On the Moon with *Apollo 17*," December 1972, and *Preliminary Science Reports for Apollos 15, 16, and 17*, SP-289, SP-315, and SP-330 (Washington, D.C.: NASA, Scientific and Technical Information Branch, 1969–73).

11 On to the Moon: Science Becomes the Focus

1. NASA memo, To: SP/Posner, From: MAL/Director Apollo Lunar Exploration Office, Subject: Questions for the Record, Subcommittee on Space Science and Applications—13 March 1969, Date: March 17, 1969.

2. Apollo Program Directive 4-K, Apollo Program Schedule and Hardware Planning Guidelines and Requirements, July 10, 1969.

3. NASA memo, To: OB/Executive Officer, From: M/Associate Administrator for Manned Space Flight, Subject: Monthly Report for the White House, with Enclosures, Date: August 23, 1969.

4. *Lunar Exploration, Strategy for Research, 1969–1975*, Report of a Study by the Space Science Board, September 1969 (Washington, D.C.: National Academy of Sciences, 1969).

5. "Opportunities for Space Flight Investigations, Follow on and Later Apollo Lunar Surface Missions 1971–1975," Memorandum Change 25, NHB 8030.1A, August 25, 1969, and changes dated September 4, 1969.

6. NASA memo, To: Manned Spacecraft Center, Attn: Col. J. A. McDivitt/PA, From: MA/Apollo Program Director, Subject: Management of Science Program, Date: October 21, 1969.

7. NASA memo, To: NASA Headquarters, Attention: Dr. R. A. Petrone, MA, From: Manager, Apollo Spacecraft Program, Subject: Management of Science Program at MSC, Date: October 21, 1969; NASA Memo originated at the Manned Spacecraft Center, To: See Attached List, From: PA/Manager, Apollo Spacecraft Program, Subject: Apollo Spacecraft Scientific Experiments Definition of Responsibilities, Date: November 7, 1969; NASA Memo, To: NASA Headquarters, Attention: Dr. R. A. Petrone, MA, From: Manager, Apollo Spacecraft Program, Subject: Management of Science Program at MSC, Date: November 7, 1969.

8. For example, *Space/Aeronautics,* January 1970, 33–42.

9. NASA memo, To: MSFC, Attention: Dr. William Lucas, Director of Program Development, From: MAL/Director Apollo Lunar Exploration Office, Subject: Advanced Development Tasks in Support of the Dual Mode Roving Vehicle, Date: July 7, 1969.

10. NASA memo, To: MA/Apollo Program Director, From: MAL/Director, Apollo Lunar Exploration Office, Subject: Documentation of Apollo 12 Samples, Date: February 25, 1970.

11. Letter addressed to Dr. Robert W. Schmieder, University of California, Lawrence Radiation Laboratory, Berkeley, California 94720, January 22, 1970, signed by Rocco A. Petrone, Apollo Program Manager.

12. Included in NASA "Twenty-third Semi-annual Report to Congress," chapter 2, part D, "Apollo Lunar Surface Science Program," January 1–June 30, 1970.

13. NASA memo, To: SM/Chief, Lunar Science, From: MT-1/Donald A. Beattie, Subject: Proposal for Use of Abandoned LEM Ascent Stage in Lunar Orbit, Date: December 18, 1964.

14. Bellcomm, Inc., letter addressed to Dr. G. Latham, Chairman, Seismic Experimenters Group, Lamont-Doherty Geological Observatory, Palisades, New York 10964, signed by M. T. Yates, Lunar Exploration Department, June 18, 1969; NASA memo, To: The Record, From: MAL/Donald A. Beattie, Subject: Activities of Advanced Program Task Group, Week of 23 June, Date: June 27, 1969. Includes actions accomplished for the Change Control Board to modify the SIVB to permit impact.

15. NASA memo, To: See Attached List, From: PA/Manager, Apollo Spacecraft Program, Subject: Apollo Site Selection Board Meeting, March 6, 1970, Date: March 16, 1970, signed by James A. McDivitt. (The attached list for distribution contained 130 addressees at headquarters, Bellcomm, MSC, and KSC.)

16. Letter to Dr. Homer E. Newell, Associate Administrator for Space Science and Applications, National Aeronautics and Space Administration, Washington, D.C. 20546, March 3, 1970, signed by Harold C. Urey, Department of Chemistry, Revelle College, La Jolla, California.

17. NASA memo, To: AA/Homer E. Newell, From: MA/Apollo Program Manager, Subject: Comment on Dr. Urey's Letter on Apollo Site Selection, Date: March 18, 1970.

18. NASA memo, To: The Record, From: MAL/Program Manager, Plans and Objectives, Subject: Comments on Dr. Newell's Letter on MSC/PI meeting of February 5, 1970, Date: March 13, 1970.

19. NASA memo, To: MA/Director, Apollo Program, From: MT/Director, Advanced Manned Missions Program, Subject: Proposed Memorandum of Understanding on Advanced Lunar Missions Planning, Date: March 23, 1970.

20. "Summary Report on the Lunar Exploration Program In Support of the Manned Planetary Mission Requirements Study," Apollo Lunar Exploration Office, NASA Headquarters, Washington, D.C., July 15, 1970, cover memo signed by Lee R. Scherer, Director, Apollo Lunar Exploration Office.

12   The J Missions: We Almost Achieve Our Early Dreams

1. NASA memo, To: M-1/Executive Assistant to the Associate Administrator, From: MAL/Manager, Apollo Surface Experiments, Subject: Television Coverage of Apollo 15, Date: June 7, 1971.

2. *Preliminary Science Report for Apollo 15,* SP-289 (Washington, D.C.: NASA, Scientific and Technical Information Branch, 1972), 4-2.

3. Eventually the decision was made not to fire the fourth mortar, so it remains on the lunar surface. Future explorers beware!

4. NASA memo, To: Distribution, From: MA/Apollo Program Office, Subject: Apollo 12 ALSEP Operation, Date: November 13, 1972.

5. "Apollo Program Plan," Office of Manned Space Flight, December 12, 1972, NASA, Washington, D.C.

13   The Legacy of Apollo

1. This program operating plan, dated May 23, 1972, just six months before the final Apollo flight, was found by a former budget colleague, Alex Schwartzkopf, on his dusty library shelves. It may well be the only hard copy in existence of this twenty-seven-year-old document.

2. NASA memo, To: M-N/Public Affairs Officer, Office of Manned Space Flight, From: MAP/Director, Program Budget and Control, Apollo Program Office, Subject: Apollo 15 Experiment Costs, Date: July 13, 1971.

3. NASA memo, To: MBD/Deputy Director, Budget and Program Analysis, From: MAP/Director, Apollo Program Budget and Control, Subject: Apollo 16 and 17 Mission Costs, Date: March 6, 1972.

4. William David Compton, *Where No Man Has Gone Before: A History of Apollo Lunar Exploration Missions,* NASA History Series SP-4214 (Washington, D.C.: NASA, Scientific and Technical Information Division, 1989).

5. Leon J. Kosofsky and Farouk El Baz, "The Moon as Viewed by Lunar Orbiter," NASA SP-200, 1970; David E. Bowker and J. Kendrick Hughes, "Lunar Orbiter Photographic Atlas of the Moon," NASA SP-206, 1971.

6. See, for example, B. M. French, "Twenty-five Years of the Impact-Volcanic Controversy: Is There Anything New under the Sun or Inside the Earth?" *Eos (Transactions of the American Geophysical Union)* 71 (1990): 411–14, and R. A. F. Grieve and L. J. Desonen, "The Terrestrial Impact Cratering Record," *Tectonophysics* 216 (1992): 1–30, for discussions relating to the implications of impacts on the Moon and Earth.

7. L. W. Alvarez, W. Alvarez, F. Asaro, and H. V. Michel, "Extraterrestrial Cause for the Cretaceous-Tertiary Extinction," *Science* 208 (1980): 1095–1108.

8. The quotation is from the *Apollo 8* air-to-ground transmission as reported in Andrew Chaikin, *A Man on the Moon: The Voyages of the Apollo Astronauts* (New York: Penguin Books, 1998).

9. "Apollo 11 Photographic and Scientific Debriefing, Operational Photography, Lunar-Surface Photography, General Observations, August 12, 1969," prepared by Lunar Surface Operations Planning Office, Lunar Surface Project Office, MSC, Houston, Texas.

10. Apollo 11 Mission Director's Briefing for News Media, George H. Hage, Apollo Mission Director, Capt. Chester M. Lee, USN (ret.), Assistant Mission Director, Col. Thomas H. McMullen, USAF, Assistant Mission Director, Back Contamination Briefing, Col. John E. Pickering, USAF, Director of Lunar Receiving Operations, NASA-OMSF, June 16, 1969.

11. Quotation contained in a newspaper clipping; unfortunately the source and date are missing.

12. C. A. Benschotter et al., "Apollo 11: Exposure of Lower Animals to Lunar Material," *Science* 169 (July 31, 1970): 470–72.

13. Lunar Sample Analysis Planning Team, "Sample Information Summary—Number 5 Final," compiled by Gene Simmons, NASA Internal Report, August 27, 1969.

14. Lunar Sample Preliminary Examination Team, "Preliminary Examination of Lunar Samples from Apollo 11," *Science* 165 (September 19, 1969): 1211–27.

15. H. H. Schmitt, G. Lofgren, G. A. Swann, and G. Simmons, "The Apollo 11 Samples: Introduction," in *Proceedings of the Apollo 11 Lunar Science Conference,* ed. A. A. Levinson (New York: Pergamon Press, 1970).

16. R. D. Johnson and C. C. Davis, "Total Organic Carbon in the Apollo 11 Lunar Samples," in *Proceedings of the Apollo 11 Lunar Science Conference,* ed. A. A. Levinson (New York: Pergamon Press, 1970), 2:1805–12.

17. F. Buhler, P. Eberhardt, J. Geiss, J. Meister, and P. Signer, "First Results [Solar Wind Composition Experiment]," *Science* 166 (December 19, 1969): 1502–3.

18. *Preliminary Science Report for Apollo 16,* NASA SP-315 (Washington, D.C.: NASA Scientific and Technical Information Branch, 1972).

19. Ibid., and selected quotations from *Preliminary Science Reports for Apollos 11, 12, 14, and 15,* SP-289, SP-315, SP-330 (Washington, D.C.: NASA Scientific and Technical Information Branch, 1969–73).

20. Graham Ryder, *Catalog of Apollo 15 Rocks,* part 2, Curatorial Branch Publication 72 (Houston: NASA, Lyndon B. Johnson Space Center, 1985), 591.

21. NASA memo, To: Distribution, From: MA/Apollo Program Director, Subject: Background Information Covering Report of Dr. Freeman, Rice University, of Finding of Water upon the Moon, Date: October 22, 1971.

22. *Science* 281 (September 4, 1998): 1475–1500 (a series of seven papers discussing various results from Lunar Prospector); *Aviation Week and Space Technology,* March 23, 1999, 80 (based on findings presented at the Thirtieth Lunar and Planetary Conference in Houston).

23. Letter from Audouin Dollfus, Observatoire de Paris, Section d'Astrophysique, addressed to [D]onald Beattie, NASA Headquarters, Code MAL, Washington 25, D.C., April 25, 1972.

24. Boris Voishol, "Lunar Geology," *Geotimes,* September 1968 (letter to the editor).

25. Bill Keller, "Eclipsed," *New York Times Magazine,* June 27, 1999, 30.

Selected Bibliography

Brooks, Courtney G., James M. Grimwood, and Loyd S. Swenson Jr. *Chariots for Apollo: A History of Manned Lunar Spacecraft.* NASA History Series, SP-4205. Washington, D.C.: NASA, Scientific and Technical Information Branch, 1979.

Chaiken, Andrew. *A Man on the Moon: The Voyages of the Apollo Astronauts.* New York: Penguin Books, 1998.

Compton, William David. *Where No Man Has Gone Before: A History of Apollo Lunar Exploration Missions.* NASA History Series, SP-4214. Washington, D.C.: NASA, Scientific and Technical Information Division, 1989.

Ezell, Linda Neuman. *NASA Historical Data Book.* Vol. 2. *Programs and Projects, 1958–1968.* NASA SP-4012. Washington, D.C.: NASA, Scientific and Technical Information Division, 1988.

Levine, Arnold S. *Managing NASA in the Apollo Era.* NASA History Series, SP-4102. Washington, D.C.: NASA, Scientific and Technical Information Branch, 1982.

Proceedings of the Apollo 11 Lunar Science Conference. Ed. A. A. Levinson. 3 vols. New York: Pergamon Press, 1970.

Science 167, 3918 (1970): 1–792. The entire issue was devoted to the results of the First Lunar Science Conference held at the Lunar Science Institute in Houston.

Wilhelms, Don E. *To a Rocky Moon: A Geologist's History of Lunar Exploration.* Tucson: University of Arizona Press, 1993.

Index